Morris William Travers—
A Lifetime of Achievement

Morris William Travers— A Lifetime of Achievement

❖

Keith Kostecka

To order additional copies of this book, contact:
Xlibris Corporation
1-888-795-4274
www.Xlibris.com
Orders@Xlibris.com
90875

CONTENTS

In writing any book, there are always many who are helpful, if not vital, to its preparation and completion. I cannot name all but there are several individuals who I wish to publicly acknowledge and they are: Heidi Marshall, Head of College Archives and Digital Collections at Columbia College Chicago, who helped locate the Travers family in the USA and in digitizing print material on Professor Travers; Dr. David Travers and the Travers family for their assistance and materials provided and most importantly, to my wife Dolores for her support, patience and love throughout my efforts in completing this biography.

CHAPTER I

A History of the
Travers Name and Family

The Beginnings

According to Ronald D. Travis, who was researching his direct ancestor "Travers", a man named Travers was a chief in the Norman army and came to England with William the Conqueror. This Travers is said to have participated in the Battle of Hastings on October 14, 1066.

Later, in another battle, this Travers and his men took the towers of Tulketh Castle and Travers proceeded to marry Alison. There is a reference found to Mount Travers and it is thought that this is noting the high ground of Tulket and the manor house of Tulketh Hall, near Preston, where the castle itself existed [it was unfortunately demolished in the second half of the 20th century].

This author discovered no information dealing with the name Travers until mentions made in the 17th century. It is apparent that Travers families lived for several hundred years at Nateby Hall until it was sold in 1626.

In America

At this time, one member of the Travers family, Edward—son of William, left for America and settled at Jamestown, VA. Edward did quite well—purchasing over three hundred acres of land near Jamestown in 1637. Ultimately, an 837-acre plantation was built up including a large number of slaves, horses, livestock, etc. A large mansion was built on this land; it though burned down in 1822 and descendants of the family sold the plantation in 1831.

Interestingly enough, both in America and England, the name Travers has had multiple additional spellings, including: Travis; Traverse and the little seen Treeves.

The Travises of Jamestown and Williamsburg were much involved with and instrumental in the beginnings of the USA. Three generations of this family were involved in government as burgesses in Virginia's House of Burgesses and one (William in 1679)—was actually the Speaker of this body! Another Travers was sent as a delegate to the convention to officially declare independence from the King of England! Of further note is the fact that the earliest Travis house (built in 1765) is still standing today in Williamsburg. It has been restored several times and has been kept as it originally was. Other family "members" who settled in America included: Walter Travis, who came to America in 1637; John Traviss in Maryland in 1734 and Joseph Travis also in Maryland in 1738.

In England

With the assistance of church registers, the family of Professor Morris William Travers can be traced back to a Richard Traverse whose sons William and John were baptized in 1624 and 1628, respectively. John had two sons; Francis who was born in 1671 and William who was born in 1673.

Records next indicate a Richard Traverse (1702-1780) and his wife Edith (1705-1785) had seven children and that Richard was the Overseer of the Poor. In 1754, he took out a lease on Bidlake, a farm in Netherbury Paris, which he and his descendants held for a hundred years. Richard's third son was John (1735-1805) but little is known of him other than his tenth child being William (1776-1854) who was the great-grandfather of Professor Travers.

William's life came just within the time periods of several individuals known by Professor Travers. The brother of Professor Travers, Wilfred, was often, while a boy, a guest of William's daughter, Mrs. Randell of Swanage, and he recalled stories told of William Traverse.

William first worked on the family farm Bidlake. He married a woman from the adjacent village of Loders in 1797 and this union led to three daughters. However, Williams's wife died in 1803. He remarried in 1806 taking Harriet Rogers of Oxford as his second wife. They went to live at Salcombe Regis.

At about this time, William was a warrant officer in the Royal Navy. By 1811 he was in command of one of a group of vessels described as King's Yachts, which were small ships carrying 8-10 guns each. Typically, these unnamed vessels served as tenders in the great ports of England; however William's ship was part of a group carrying out secret missions as directed by the commissioners for the Navy.

It is possible that William was involved in an interesting trade between two nations at war: England and France. The English wanted claret—it was the drink of their gentry and, particularly of the court and the French wanted coffee. This trade was not permitted given Napoleon's continental system aimed at sterilizing English sea-borne trade and the English response to France was obviously not favorable to say the least!

Apparently William was involved in the smuggling of claret for the English gentry from France. In fact, it has been said that the French armies were financed to no small extent by English gold! In 1811, William was taken prisoner in Nantes in 1811 while negotiating a deal in claret while he was likely the ship's purser. Interestingly enough, though acting as a smuggler, William was well treated as a prisoner of war [particularly by the Governor of the facility who was a Freemason and liberal in his views] while in Verdun.

Upon his return to England in 1814 after the war was over, he was far from poor and no doubt had benefited from trading in gold. William then stayed for a time in Corfe with his son John and then settled in Swanage where he died in 1854. His sons, except John who eventually went to Martinique, did well and his daughters married well.

His youngest son Richard (1821-1904) joined the staff of the National Provincial Bank in 1842 and became Manager at Bishop Auckland in 1854 eventually becoming senior manager until his retirement in 1878.

William's second son Frederick (1808-1876), grandfather to Morris William Travers, was educated at Oxford. He became an assistant to William Curtis at Abingdon who was head of a school for young gentlemen. He later married the daughter of his employer in 1835 and moved to Poole where he opened and served as headmaster of the West Street Academy until his death in 1876. His contemporaries, particularly William Whewell, Master of Trinity College in Cambridge, considered Frederick a scientist.

A collection of Frederick's beautifully constructed electrical apparatus was passed on to his grandsons. In addition, he paid considerable attention to the improvement of Poole Harbor and studied the tides to help Sir John Coode who was employed to advise on improving the entrance to the harbor. Frederick was also a sound classical scholar and had marked literary ability as noted by his being asked [though he declined due to his disgust with the nature of advertisement] to contribute examples of English handwriting to the 1851 exhibition.

Frederick Travers had four sons one of whom was also named Frederick. This younger Frederick, the oldest of the four, had a distinguished career with the National Provincial Bank.

William Travers (1838-1906)

The father of Morris William Travers was the second son of Frederick Travers and was born, like his older brother, at Abingdon. William was educated under the direction of his father at the West Street Academy. He was later a pupil of Dr. Thomas Salter of Poole who was the first to use chloroform in surgery and possessed a private anatomical museum which, upon his death, became part of the Dorchester Museum. Dr. Salter's sons and brother-in-law all became Fellows of the Royal College of Surgeons (F.R.C.S.).

Dr. Salter died before William could finish his medical training under his direction. Dr. William Travers later though qualified for his profession at the age of 21 and then began working at the Charing Cross Hospital.

William made rapid progress there and became House Surgeon by 1859. In 1861 he succeeded Dr. Benjamin Golding (1793-1863), founder of the hospital, as the Resident Medical Officer. William held this role for five years until 1866. He was the doctor who dealt with all the urgent cases when the honorary staff left the hospital at the end of the day—even if it called for major operations to be performed with limited use of anesthetics, no use of antiseptics and poor nursing. He himself became a member of the F.R.C.S. in 1864.

He began his private practice in Kensington in 1866 and remained at work there until his death in December 1906. William also served as Physician to the Chelsea Hospital for Women from 1883 to 1894. He was also an early disciple of Joseph Lister and was one of the first practitioners of aseptic surgery. A comment made by William to his eldest son Fred (born in 1869) during Fred's medical education stands out; "I have never lost a mother". This spirit and accomplishment was memorialized in the William Travers Prize in Obstetrics and Gynecology from the Charing Cross Hospital; this prize is still awarded today in 2011.

William was a founding member of the Gynecological Society in England. He was a member of its council for many years after its inauguration and for a few years its honorary treasurer. He was not able though, due to ill health later in his life, to accept the Presidency of this society when offered. Other organizations that he was a member of included the: British Medical Association; Medical Society of London; Clinical Society; West London Medico-Chirurgical Society (of which he had been President) and the Anthropological Society. William was also a Freemason (since 1865) and was a past Master and Father of St. Mary's Lodge. He also was one of the founders of the Durham University Lodge and of the Cavendish Chapter.

Anne Travers, nee Pocock (1843-1923)

The first Pocock ancestor found by this author was the Reverend Thomas Pocock, (F.R.S.), whose sons were Admiral Sir George Pocock K.B. (1706-1792) and the Reverend John Pocock.

Reverend John Pocock's grandson was John Thomas Pocock (1770-1832), of St. Bride's Wharf, who bought property in Islington that became known as Pocock's Fields. He was a merchant who owned several coal mines and their associated buildings in South Wales. His wife Margaret was the daughter of George Kennedy of Wigtonshire who was significantly involved in the development of local agriculture and horticulture. In fact, Margaret's brother became the chief horticulturist in France under Napoleon. Margaret died of tubercular trouble that came about after catching a chill in the aftermath of being outside and poorly dressed after a fire destroyed the St. Bride's Wharf in 1809. John Thomas Pocock outlived her by 23 years only to die in a cholera epidemic of 1832.

Samuel, the eldest son, inherited their property while their third son, Lewis, gained enough distinction as an art critic and writer to be listed in the Dictionary of National Biography.

Second son Thomas (1805-1869) was educated as a lawyer and inherited a share of his father's wealth and his uncle's house and legal practice. He was married twice, the second time to Caroline Crossthwaite and this union produced four children, including Anne Pocock who was their second daughter and third child overall. All of Thomas and Caroline's children were born and brought up in the city though they also had a home in the Notting Hill district where they lived in the summer.

Thomas, besides being a lawyer, was also a financier who built many homes in the Ladbroke area including a home for he and his family at the northeast corner of Ladbroke Square [which later became the location for the Japanese embassy].

Marriage of Anne and William [and Tragedy]

On January 28th, 1869, Anne Pocock and William Travers were married at St. John's Church [Kensington Park Gardens], which happened to be just a few yards away from Thomas Pocock's home at the corner of Ladbroke Gardens.

After their wedding, Anne and William went to Oxford; but they had to return two days later. Thomas Pocock had slipped on the icy Exeter Station platform on January 26th; a hernia he had previously became strangulated. Even with the best efforts of Dr. James Paget, Mr. Pocock died.

Mr. Pocock had been engaged in a number of speculative enterprises, one of which was an attempt to recover Chichester Harbor. His involvement was found to be significant in this venture and lawsuits absorbed much of what was left of his money. In fact, Anne was to have received the amount of £1,000, with which she intended to buy out her now deceased father's partner. To buy him out, this sum had to be borrowed and it proved to significantly alter the lives of Anne and William.

Given this financial crisis, Anne became William's assistant [this "arrangement" lasted for ten years] and she had to learn, at once, the art of making up and dispensing medicines for his patients. Even when William was making £4,000 per year, Anne still did the accounting work needed for his practice as well as tending to the social side of William's medical career.

CHAPTER II

Childhood and Youth

Childhood

Fred, the first child of Anne and William, was born on December 3rd, 1869. A daughter was born in November 1870 but unfortunately died about a year later.

Morris William Travers was born on January 24th, 1872 but was in poor health. His health was so delicate that his father William had to call on the famous child's specialist Dr. Charles West [the founder of the Great Ormond Street Hospital in 1852]. Dr. West's efforts were successful and Morris "pulled through". West, who did much to improve the medical care of children in England in the 19th century, took no fee from William Travers.

William Travers set up his medical practice at 19 Lower Phillimore Place on the Kensington High Street. This row of homes was built in the last third of the 18th century when the Court was still centered on Kensington Palace and thus a host of minor officials were in the district. Looped strings in plaster on the fronts of these homes in fact earned them the name "Dish Clout Terrace" from none other than King George III. But they were more than just that—they were well-arranged buildings with sound internal construction, plain moldings and the simple, yet well-designed fire grates of the period, which kept the fire well above floor level.

In June 1873, the Travers family left this address for a larger home at 2 Phillimore Gardens. This move was finished just a few days before the birth of Morris's brother Harold who was named after William's friend Sir James Paget. The new Travers home bordered on Holland Lane which was an avenue having many elm trees and a large rookery. From the back windows of their home, the Travers family looked out over Holland Park with Holland House away to the right and the Kensington Main Road, which was almost hidden by trees, to the left.

Harold Travers was born a few days after the family moved into this home. Cecil was born on January 26th, 1876 thus giving Anne and William a total of five children with the oldest being just over six years of age at that point in time.

All five children were left in the care of their nanny, Caroline Johnson, who remained with the family until 1879. She was particularly fond of Morris and evidently had much to do with improving his health. In addition to Ms. Johnson, the Travers family also employed a succession of young girls as assistant nurses, three maids and a boy who came in to deliver medicines to William's patients and to clean boots and knives.

Up until 1878, the pantry, which was behind the kitchen, had also been the dispensary for medicines. This changed when patients began to take their prescriptions to their local chemist for filling. In that year, the Travers family added a butler to the establishment. The first one was a faithful Cornishman named Penderick who went with the family when they went on vacation.

Youth

The life of the Travers children was simple. It involved the routine of getting up in the morning and eating breakfast where simple, reasonable rules were followed. After breakfast came the morning walk, when the Travers children met their friends in Kensington Gardens where they played games and fed the ducks on the Round Pond. Afternoon walks were similarly uneventful; but sometimes, in the summer, the children took their tea with them to Kensington Gardens.

But these pleasures came to an end with the hiring of nurses for the children. Meals were very simple. Breakfast consisted of cereal, milk, and bread and butter, typically cut from large household loafs; and dinner of a roast or a stew [often Irish stew], and generally a milky pudding and pastries. Tea always began with bread and butter, followed by cake or bread and jam, but never with butter under the jam, and never by both unless there were guests at tea. Milk, a biscuit, and perhaps a piece of chocolate ended the day. This was quite typical of the feeding of the middle class children from this era; the almost complete lack of fruit in the children's diet was based on contemporary ignorance. In general, this diet overall for the Travers children was an improvement on the childhood diet of their mother Anne and even that of her very well-to-do father Thomas Pocock.

Pocket money for the children was very limited and for the children to ask for money was just not done. Allowances were typically no more than two pennies per week until a child reached ten years of age when they would receive four pennies per week. Presents were given at Christmas and on birthdays. Toys were gradually supplanted, as the children grew older, with books, tools and the like. On occasion extra presents gave great delight to the children; many of them were shared, such as the large rocking horse in the nursery, on which the children tried to imitate the feats [hopefully without injuring themselves] seen at the circus. The Travers children only knew theater from their annual visits to Drury Lane at Christmas.

Anne and William were always very generous to their children when it came to country vacations and Morris in particular recalled them, in his early years, as happy and uneventful. Liking the water was something that the Travers children did with ease; Morris felt, after the age of seven, that the sea meant freedom. In fact, many years later, Morris's American daughter-in-law and her children came to stay with Morris and his wife in the country. His daughter-in-law was worried that if her children became wet, what would they do? Morris's experiences as a child led him to say that, "If we got wet, we just got dry again".

Children's parties for Morris and his siblings at the time of Christmas were occasions of splendor. They would typically include a "magic lantern", a magician, a Christmas tree or something of the kind, and, at first, a wonderful sit-down tea which likely merged into a sit-down supper, and ultimately into more or less grown-up affairs with continuous refreshments and champagne throughout the day with supper a la fourchette. Ice creams were made from cream or water ices, and boned turkey and other substantial food was typically on the menu with soup appearing early in the morning. Anne Travers was masterful at putting this magnificent feast together while William's dance champagne was quite good.

The change from infancy to childhood in the Travers family was a sudden process for the individual but a gradual one for the group. It started in 1877 with Fred being sent in the mornings to a day school located over a shop adjacent to the Addison Road Station. Then, he was sent to a boarding school at Hastings. Morris remembered little about his brother from this institution at all but he did recall seeing the school's swimming pool, having tea with the Headmaster and a young lady showing him aspects of archery that seventy years later he was able to use in teaching his American grandsons while making bows and arrows for them.

The governess era soon began for Morris and Harold. This brought about the bad and the good. One governess used to lock up the boys when they misbehaved in a dark cupboard as a punishment. This practice continued until she was caught in the act and fired immediately by Anne.

Morris began learning the piano at about the age of five with his governess sitting beside him and correcting his mistakes with a ruler to the hands when he made an error. He learned the major and minor scales and various piano pieces including "God Save the Queen" and the hymn "My God, my Father . . .". This introduction to music continued until he was seven when a change of governess occurred. Later in life, he regretted dropping music since his mother and uncles were talented musicians and he married a musically talented woman.

Morris noticed about this time that for the three elder boys (Fred, Morris and Harold), he always seemed to be left out of the "picture" that included Fred and Harold. This later in life though turned out to be a plus where Morris and Anne remained close enough that she always turned to him when anything unusual, even unpleasant, had to be dealt with.

All in all, the time spent under the direction of three different governesses was valuable. Morris memorized a vast amount of verse, and thoroughly read Dr. Brewer's Child's Guide to Knowledge, which he loved. In fact, at one point during World War I while occupied at the glass works, Morris when asked where he got a particular piece of useful, but arcane information from, he replied "Out of the Child's Guide to Knowledge, when I was a kid".

School and Summer Vacations

Even for Morris the governess era soon came to an end. In the spring of 1879, Harold, who was supposed to be delicate, was sent to his father's cousins, the Randells, at Swanage. Morris, in turn, was off to a day school in Scarsdale Villas, Kensington. At this institution, Morris did not take his time seriously but he recalled his first evening's work very vividly. He was to read the first chapter of Dr. Brewer's History of England. This work opens with:

Q. Who conquered Britain?
A. Julius Caesar, B.C. 55 and 54.

Morris was not able to get any further. When he asked his nurse Caroline as to what B.C. meant, she, and the entire kitchen staff, could not give him an answer. Anne, who had been out, attempted to explain "B.C." to Morris without any success. He also recalled being drilled continually by a big and pleasant Sergeant of the Guards from Kensington Barracks but only on what were his right and left hands and that his right hand remained his right hand after he had turned about. In addition, Morris did remember watching construction on shops on the Abingdon Road, which quite often led to his being late for school.

The beginning of the summer of 1879 was a particularly lonely one for Morris with his elder brother Fred at school and his younger brother Harold at Swanage. With the next child in birth order being according to Morris himself, "only a girl", he was fortunately taken in by his nervous and reclusive, though kind and well meaning, father William. William took Morris, during this summer, to Greenwich on the river and to see the Panorama of the Charge of Balaclava in Trafalgar Square. William brought Morris home tools including his first carpenter's plane that Morris proceeded to use to get into trouble with his mother. The enterprising small, lonely young man had, one afternoon, taken last year's Christmas tree and set it up in the drawing room for some carpentry practice. Anne was not pleased at seeing this woodworking in progress and was about to inflict punishment on Morris until William arrived to set matters right.

Morris knew little of his father's professional activities at this time in his life. It turned out though that the summer of 1879 was a particularly busy one for William; he had decided to take an M.D. at Durham University. This required passing a matriculation including: French, Latin and Greek. In achieving this mastery, William had to read Caesar's Gallic War in Latin and the Gospel of St. John in Greek. Accomplishing this at the age of 41 with a large, young family around was quite remarkable!

Summer vacations before 1879 were not well remembered at all by Morris. The first vacation that he had any recollection of was spent at Littlehampton. All he remembered of this was of being sent out to play on the beach while his mother and the nurse did the unpacking. He tried to find his way home alone but got lost and went to the wrong house.

The summer vacation of 1879 though stood out well in Morris's memory. His family went to Swanage [where Harold was] and while going up there in a small, private bus had Fred, who was riding on its step, fall off when they were near Corfe Castle. He got up red from head to foot, much to the alarm of Anne, but this was just the red dust of the area and not blood.

Harold, who had a very outgoing and engaging personality, had been in Swanage for two months and knew everyone in the village.

Interestingly enough, there was a flourishing building and paving stone trade at Swanage that was partially owned by the Travers's aunt Mrs. Randell, who had taken over her husband's business after his death years back. The stone was quarried and cleaned at the quarry pits on the hill overlooking the village, and stacked on platforms along the shore. Then, it was taken in carts to what were called "stone boats" and loaded from them into vessels with two square-rigged masts ("brigs") to be carried to London.

Some of the stone was carried on rails to a wooden jetty for loading. Fred, Morris and Harold, not being controlled by governesses or nurses, had a great time climbing up and down the stone piles on the platforms and sometimes being taken out to the stone boats while in the carts. The boys also spent time with their butler Penderick by going fishing off the jetty, boating in the bay and sailing around the coast.

The autumn of 1879 saw Morris at school at Augusta House, Ramsgate, under the care of Miss Elgar and Miss Tate. These two able women, while good instructors, had little idea how to care properly for the health of children. Ill health to them was due to original sin and was to be corrected only by punishment. Morris spent three years at this school and had his summer vacations at Hythe in Kent where he [and his brothers] swam, rambled and fished in the Royal Military Canal. He wrote a letter to his father on William's birthday in August 1880 where he told him that he had a fishing rod made from a bull-rush stem that had allowed him to catch three fish. These were the first of the hundreds of fish that Morris caught over the next 60-70 years. Other than the fishing, this vacation was rather dull. It involved playing cricket [never Morris's favorite] with a neighbor family and catching dormice with Harold. These small creatures along with horses were the only pets Morris had during his life; he hated highly bred fancy dogs, which he said were too tame.

Morris left Ramsgate at the end of the summer term in 1882 and then spent a month with his family at Sandown Bay. Swimming and digging on the beach occupied most of this time for Morris. He also, during this vacation, had the chance to see the H.M.S. Thunderer, England's first turret ship, and its 81-ton gun that had burst. A visit to Portsmouth was also made to see the H.M.S Victory [launched in

1765 and Nelson's ship at the Battle of Trafalgar]. In 1882 this vessel was still afloat; in fact in 2011 it is still a commissioned vessel in British navy.

Following this visit, Morris and his brother Fred spent two weeks with Morris's godfather, Evan Morris, at Wrexham. The future Sir Evan Morris had a nice home and garden but his family consisted of six girls with the oldest Daisy being a bit older than Morris. The Travers brothers did not fit in well with this family and spent part of their time in smoking out wasp's nests, getting stung by wasps and damaging cucumbers where the wasps lived.

When the vacation was complete, Anne Travers, with a War Office permit, took Fred and Morris to see the Woolwich Arsenal [which was a facility that carried out armaments manufacture, ammunition proofing and explosives research for the British armed forces]. Morris did not revisit this site until 1940 when he went as a Consultant to the Ministry of Supply.

September 1882 had Morris and Harold sent to Woking College, a large white house at the entry to Old Woking Village, under the direction of Reverend C.W. Arnold, M.A. This man, who had shamefully left a school in Devonshire due to his excessive severity, was, according to Morris, a person of keen and irritable temper. Reverend Arnold's rule of life was work; and as to hours of sleep he allowed "six for a man, seven for a woman, eight for a child and nine for a pig". As if this needed any proof, any boy that stayed in bed after 6:30 AM was likely to have his pajamas pulled off and receive a blow with the old man's cane.

Morris regarded Woking College as a rough school and he was certain that other schools had rejected many of its boys, ranging in age from 8 to 18. The masters of the school were also equally rough and the little boys had a terrible time. Morris felt that, in spite of all this, Arnold meant to be kind in his own way, and he certainly fed his students well. However, due to his experiences in spending many weary hours as a virtual slave to the older boys who were practicing cricket, Morris developed a hatred for the game. In fact, William Travers said years later to Harold that he regretted sending Morris and Harold to Woking College. Interestingly enough, Arnold gave up the school a few years after Morris and Harold left and went to California in the USA to grow fruit.

In the spring of 1885, Morris went to Blundell's School in Tiverton. Blundell's was then developing into a public school and had a good chemistry lab facility.

Morris's brother Fred had been there since the autumn of 1883 (having had his troubles previously at Epsom College). The house that Morris was in during his early time at Blundell's was a happy, but rough one. However, that now began to change due to the neurological trouble [from which he died in the autumn of 1889 after Morris had left] that the

housemaster was experiencing; he had no control over the house but he would occasionally descend upon a near riot at the house and inflict savage punishment on all. In addition, his wife took no interest at all in the house and those in the house were near starvation.

Success on the academic end also proved difficult for Morris. He entered in Form IIIB. This class was revising two terms of work on syntax from the public school Latin Primer, which had been learned by heart by these students. Morris had never used this book before and the form master did not care and Morris was frequently beaten for failing tasks he did not have the needed background to accomplish. Morris remained in this class until Christmas, when he was promoted to Form IIIA; and coming into his first course of chemistry a term behind the rest of the class, he won the chemistry prize in the following June, when he was promoted to Form IV (modern), and a year later to Form V (modern).

This was a small form of five boys and Morris ran into the same ill luck as he had in Form IIIB. He found this class to be revising Cicero's *De Senectute* [a work on attitudes toward aging] and that he was expected to prepare two chapters for each of the two lessons in the week, which for him, was impossible. However, in the following June Morris won the form prize and the chemistry prize. But this left him no fresh fields to conquer since there was no Form VI (modern). Morris firmly believed that the headmaster refused to have Form VI boys who would have been out of direct contact with him; which would have been so if there were a Form VI (modern).

The vacation of 1885 at Aldeburgh was one that Morris himself recorded information on. Fred, Morris and Harold were keen insect catchers but not entomologists at all. They each bought death's head hawk moth caterpillars [which fed on potato haulms] from a villager. The three brothers provided boxes filled with soil for the animals to burrow into when they were ready to pupate. However, no moths came about from this venture.

This vacation was also the first one that the brothers had their own boat. Though Fred was fifteen and the youngest of the children was not yet eleven, Anne Travers had the governess of Morris's sister, who was according to Morris a charming young woman, be the captain of the boat. No undue risks were taken with the tide and no time was spent stuck on mud banks. The children had a lot to learn about boating; these lessons later served Morris well in life when piloting a small boat while fishing in Scotland or in Norway, or in meeting sudden unexpected storms.

An uncomfortable vacation in Dovercourt was relieved by William's gift of a tandem tricycle to Morris. Unfortunately, it had a habit of dropping balls from its bearings, and one could not travel far with it. In 1887, William gave bicycles to Fred and Morris. These also saw more time in the repair shop than in use. William's intent though led many of the Travers children to enjoy the bicycle age of the 1890's. Weekly recreations were often a fifty-mile "ride" out to Purley or St. Albans and then back with a sandwich lunch and tea at a cottage along the way.

After Morris had registered to enter University College in London in June 1889, the form-master asked him how long he had been in the class. Morris answered "Over three years, sir" to which the master replied "Impossible". But this was easily confirmed when the master checked Morris's enrollment status and found it to be true. Nothing could have been more discouraging than seeing one's juniors on the classical side pass in succession over one's own head. When Morris left at the end of the term his report was marked "Promoted into Form VI" where the words were doubly underlined; though that was of little use to Morris.

The last of the summer vacations that the family as a whole [except for Fred] participated in were from 1888 to 1891 at Swanage. Much of Morris's time in good weather was spent in boating on the bay and outside of it, and in catching lobsters in wire netting pots. There was also a good deal of open country around them at Swanage, which allowed for walking along the coast to St. Alban's Head in one direction or to Studland Heath in the other.

As a final remembrance of the days of his youth, Morris Travers, in January 1954, while the Chairman of the Old Boys Dinner in London for Blundell's School, gave a speech that read:

> "It will be seventy years ago in the middle of April next since I entered Blundell's School. Strangely enough, of the few memories which remain of that first term one is of the Old Boy's luncheon which was held in the middle of June, which was, I suppose a survival from the time when the school year was divided into halves, instead of three terms. We boys were admitted to listen to the speeches; and to pick up such items of dessert as friends at the table might pass out to us. I remember that a very old man, as he seemed to us, then, spoke of a night in June, 1815, when the Headmaster came into the dormitory and told the boys that the Duke of Wellington had won a great victory over the French. The boys got up in bed and cheered. So I, and the other Old Boy, can carry you back 140 years; I and three others could carry you back to the days of John Ridd and Lorna Doone, and five of us to the foundation of the school in 1604.
>
> "Fragments of school history have been gathered by diligent and enthusiastic Blundellians; but I wonder if we really know anything at all about life in the school more than a hundred years ago. Blackmore has given us what many have assumed to be a glimpse of John Ridd at school at the end of the seventeenth century; but he wrote less than a century ago, and the fragment which he gives us has a suspicious resemblance to an incident in Tom Brown's School Days, written a little earlier. So far as I have read social history, it seems to deal with the lives of grown-ups, and the conditions under which the children really lived has failed to survive. I can only suggest that, the weaklings having been automatically weeded

out, the survivors must have had an even larger share of original sin than those who came into the school stories of my own time, and into my own experiences. They had quite enough of it to make life interesting.

"In my time the three chief Devon schools, were Blundell's, the Imperial Service College at Westward Ho, and the Newton Abbot College. The late Sir Arthur Quiller Couch, who was at Newton Abbot, told me that they thought Blundell's rough. Kipling, who gives a dramatized account of Westward Ho in Stalky & Co., and in his short autobiography an account of the life at the school probably much as it was in the early eighties. As I have said, he calls Blundell's 'our dearest enemy'. The schools were probably not dissimilar. You will notice that the Stalky stories cover only the period of the boys lives between thirteen and a half and fifteen and a half, when Stalky had to consider seriously the job of getting into the Army, and Beetle worked directly under the Head, with a view to journalism.

"Think how we lived in the Old House at Blundell's. We slept in dormitories holding five to a dozen boys, open to inspection at any moment; but, when the house-master was supposed to be dining out, dormitories engaged in warfare. Sometimes he turned up unexpectedly, when the last to retreat from the scrap were caught, and suffered immediate execution. No enquiry was held as to who organised the fun. The victims took their punishment with smiles, and the cost of burst pillows and torn sheets was charged to no one's account.

"The day accommodation consisted of a so-called sixth form of study, for the five senior boys. The next eight in seniority had another small room. The remaining twenty-five or so, lived in a room about twenty-five feet square, called the play room, expanding into a cloak room, and small covered yard. To describe the actions and reactions of boys averaging say fourteen and a half, one would have to carry one's self back, and try to think as they did, which is quite impossible. Groups of the community and individuals were constantly bickering, but only twice do I remember quarrels being ended in an organised fight, such as described in Lorna Doone and Tom Brown. I remember an exchange of blows with another boy, F.J. Siordet, which made both of our noses bleed, ending some weeks of mutual irritation, after which the two of us became very good friends. During the nineties my quondam opponent often dropped in at our home in London for Sunday supper, till the Boer War came, and he was killed at Pardeburg. I think I also mentioned the November 5th celebrations. Home-made fireworks—I was an adept in this branch—caused casualties; but when we were carried off a large part of a canal bridge for the bonfire, Francis thought things were going too far, and the Guy Fawkes day celebration came to an end.

"My hatred of cricket led to spending half holidays wandering over the country searching for dormice with my brother Harold, and "bug hunting" in summer time.

"A Kipling could make a story of our deeds and misdeeds. I will say one thing; life was rough, and the small boys were ill protected against some of the more ruffianly; but our misdeeds were not suitable for the Freudian studies to which some modern writers try to subject school life."

CHAPTER III

How He Became a Chemist

Contact with science

Sir William Ramsay grew up amidst a group of relatives all of whom were actively interested in some branch of science, and were ready to discuss the problems of the day. This accounted for the fact he had "chemistry in his blood".

Morris Travers likely also inherited his interest in science but by a different mechanism. His father William was a widely read individual in his youth and kept abreast of the very rapid developments in his field of medicine. His children knew William's consulting room as the study and it was lined with assorted books. Morris recalled William, on many occasions, entering or leaving the home with a bundle of medical journals under his arm, which he read in his closed horse-drawn carriage between visits to his patients. Work of any kind was dropped when it was mealtime; though as he left the room after dinner, William often did say, "I really must read some this evening".

Morris's first contact with science was through a visit to the Regent Street Polytechnic with his grandmother in 1880. He was very interested in what he saw and he remembered much of that day. The diving bell where visitors could descend into a tank of water fascinated Morris. However, an angry Morris was not allowed to try this. He though remembered clearly a man with a brown beard standing behind a long table and conducting chemical experiments, including: the reaction of phosphorus and iodine—as related to matches and the preparation of a liniment. These activities thrilled Morris as he wondered why the iodine used was a solid. It would though be a while before he saw his next chemical experiment.

His next contact with science was in January 1882 when William, taking a whole day away from his practice, took Morris to the Electrical Exhibition at the Crystal Palace. The stalls were lit up by Edison-Swan, carbon-filament glow lamps. A leading London newspaper of the time said—"Someday, rich people might use this wonderful invention to light their reception rooms, but it could never come into general use." They were completely wrong though as was seen at the Health Exhibition at South Kensington about three years later where the average home was lit by electricity while the non-typical home was lit by gas.

While at the Exhibition Hall, William bought a set of Gëissler tubes for the boys. Little did Morris know that a Plücker tube, the first and simplest form of the Gëissler tube, would play such a critical part in Morris's scientific research of the future. A cylindrical electrical machine that had been made by William's father was re-conditioned and later gave a good deal of enjoyment to the children. The Travers men stayed late that night to see the place lit up by an arc light operated by a new invention—the Brush Dynamo.

A short time later Fred made a small induction coil from components sewn to a card with printed instructions costing one shilling. Fred then bought a Bunsen battery (a carbon primary cell composed of a zinc anode in dilute sulfuric acid separated by a porous pot from a carbon cathode in nitric acid). The Gëissler tubes, induction coil and Bunsen battery gave all the boys some amusement but they were not at all careful with the sulfuric and nitric acids. Damage to clothes and carpets soon called for strict rules; and Morris had an idea of what he was later to know as a chemist.

The Jules Verne work, Twenty Thousand Leagues Under the Sea, was a written work that further fostered Morris's interest in science. This book was originally a Christmas present to his brother Fred in 1881 but he did not care for it. Morris seized this book and devoured it even though it was supposed to be too "old" for him. This book, and others like it in the school library, helped Morris find solace through a very unhappy time at Woking from September 1883 until April 1884.

Morris's interest in science became more apparent while he was a student at Blundell's. He won, as previously noted a Form prize in chemistry and unlike the other students actually enjoyed the weekly lecture on chemistry. Three years of contact with Science Master and Head of the Modern Side G.H. Spring, who had wide general scientific knowledge and capable hands for performing successful experiments, was valuable and appreciated by Morris. A class in Morris's last year at Blundell's on practical chemistry, neither rigorous nor exciting, did not lessen his liking for the discipline.

University College—London

Two final points that led Morris to enter the University of London in October 1889 on the intermediate course for the degree of B.Sc. were: the inspiration of a relative named Vatcher who was a well-known analytical chemist and that William advised a course of study at University College London with the stipulation that Morris should read for a degree in science at the University of London. This advice from William was important since there were few, if any, scholarships available and William willingly paid all Morris's expenses until he graduated.

The university was then an examining body and the college courses were framed to meet the requirements of the examinations. Morris in fact stated that:

> "The course for a degree in science opened with what may be called an educational period, during which the student attended lectures and practical classes in chemistry, physics, zoology, botany, and pure and applied mathematics. In the final degree course there were three subjects for the Pass degree, in any one or more of which one might sit in an Honours examination a few weeks later".

This program, though giving a general knowledge in the leading branches of science, was terribly overloaded and unbalanced. Ramsay gave four lectures a week on chemistry to his general class during all three terms; but the practical work in chemistry occupied only two hours a week in only two of the three terms and was not, in Morris's opinion, very interesting. It was so bad that by the end of the year Morris had developed a dislike for chemistry. The physics lectures, given by G. Foster Cary, were also not appealing to Morris due to the Professor's inability to keep order in the classroom while the mathematics, taught by M.J.M. Hill, did a little algebra but did not at all touch on calculus.

Interestingly enough, there was a discipline that did significantly interest Morris. He was fascinated by Sir E. Ray Lenkester's three lectures a week on zoology, even though each lecture typically was an hour and half long. These lectures were balanced by four hours a week of dissection and microscope work. Morris even mused that had there been any possibility of making a living as a zoologist, and had he been good at drawing, he would have become a zoologist instead of a chemist!

In his second year at the university, Morris entered the main chemical laboratory and attended lectures by Ramsay on physical chemistry and J. Norman Collie on organic chemistry. The 1890-1891 session was only Ramsay's third at the University College since he left Bristol, and he was only in the early stages of building a research school. Given this, only Ramsay had a private laboratory while both of the Assistant Professors, J. Norman Collie and R.T. Plimpton worked in the main laboratory.

At the time Morris entered the laboratory, Collie had completed a good bit of work on the methylamines, eventually leading to the discovery of methyl fluoride, and Plimpton was working in rather a haphazard way on the action of acetylene (C_2H_2) gas on solutions of metallic salts. Linde and Picton were studying pseudo-solutions and this work had its origins in Ramsay's early work on suspensions. E.C.C. Baly was working on the pressure/volume/temperature relations of common gases and discredited the use of the McLeod gauge as a means of studying low pressures. There were also two female students working in the laboratory. One was attempting to determine the atomic weight of boron while the other was measuring the latent heats of evaporation of liquids.

This mixing of professors and students was quite helpful to the junior students who only had to turn to their neighbor for assistance. It also showed the junior students that research was something that they all could do and succeed at. Even with Ramsay himself, this spirit existed. He typically had a few words with each and every student in the laboratory every day.

Soon, Morris became interested in Plimpton's experiments with acetylene [difficult to obtain in those days] and Plimpton had him investigate its actions on solutions of mercury salts. To that point the only way to make acetylene had been to allow a Bunsen burner to "burn back" and the products of its incomplete combustion, containing acetylene, were aspirated through a solution of copper (I) chloride in ammonia. The resulting red copper (I) acetylide was precipitated out and when treated with weak acid yielded acetylene.

From a paper by the French chemist Maquenne, Morris learned that barium carbide (BaC_2) might be obtained by heating a mixture of barium oxide (BaO), carbon powder and magnesium powder. The product of this reaction gave acetylene with the addition of water.

Morris, after some thought, decided to try to make calcium carbide by heating a mixture of calcium chloride, carbon powder and sliced sodium in a steel tube that was closed at one end. The product of this inexpensive reaction was acetylene in high yield. Morris wrote up his work in a paper with the title "A method for the preparation of acetylene" and submitted it to the Chemical Society hoping they would publish it. However, their publications committee thought it was not important at all and relegated it to the Proceedings of the Chemical Society which was generally a publication thrown away by most recipients. There was also no reference, by title, to Travers's significant discovery in the Society's Abstracts.

This work though proved to be an extremely important paper in that it anticipated the work of Edward Goodrich Acheson and Henri Moissan by some months in discovering the commercial method of producing carbide in the electric furnace. Morris's discovery was a bar to patenting the action of water on calcium carbide; this process rapidly led to an important new industry. Had he patented the process, Morris would have become a very, very wealthy man! He did not do so because he felt very hurt by the treatment of his first published work by the Chemical Society. Later in his life, he was to learn that an "unusual" paper was looking to be summarily rejected.

Morris continued in this area by treating an ammonia solution of mercury (II) cyanide with acetylene. The product of this reaction was the new compound HgC_2. He and Plimpton described this work in a paper that the Chemical Society published in their transactions even though Travers and Plimpton regarded this work as relatively unimportant. Nonetheless, having learned an important lesson from their previous experience with the Chemical Society, Morris and Professor Plimpton patented the production of this compound for use as a percussion cap explosive for firearms.

In Morris's time as a student, the world of chemistry was in a state of unrest and the dominant area of chemistry was organic. Students were at that time less concerned with gaining knowledge than with passing examinations. The nature of the questions in Morris's time were determined years previously by a university senate and by examiners who were not teachers.

In fact, regulations for the degree in honors in chemistry at the university specified that the examination should be primarily in organic chemistry. Morris, having taken his B.Sc. degree in October 1893 in chemistry, physics and geology, took the exam for honors in chemistry in November 1893 and took second place.

Work with Albin Haller in Nancy

Given that the "tide" was still flowing in favor of organic chemistry, his experiences with R.T. Plimpton, the creative guidance given by J. Norman Collie and that he was advised by Ramsay to study organic chemistry in Nancy under Ramsay's friend Albin Haller (who had devised a number of interesting and ingenious syntheses), Morris decided to study in France with Haller. He also knew that it was customary at that time for students who wanted an academic career to spend two years abroad, typically at a German university, getting a Ph.D. degree by research.

Morris arrived in Nancy at the end of the spring 1894 term even though he felt that Haller was not as good as the leading organic chemists in Germany. Haller had been working on the reactions of phenyl isocyanate with hydroxy acids to generally give anilides. Morris was asked to study the action of the isocyanate on the oxyacids [citric acid as an example]. This gave a rare mixture as a product. Given this difficulty, lactic acid was suggested; on his own initiative. Morris experimented with lactic ethyl ester and began to see results.

At about the halfway point of the 1894 summer term, Haller informed Morris that he had been appointed Professor of Chemistry at the Sorbonne in Paris. From this point, Morris saw little of Haller. This led to Morris's increasing lack of interest in this study; though he later finished the work later in London, Haller thought he had dropped the work completely and was pleasantly surprised when Morris gave these results to he and his students.

Morris later noted that Haller had found him agreeable quarters with the Lecerneys who had retired from a flourishing lace business though investment losses had compelled them to come back to the lace business. Nancy also "housed" a number of British guests, students at the famous Forestry School, whose reputations were so well established that Morris, as a fellow Brit, was warmly welcomed. Morris had also joined the Cercle des Etudiants, which had quarters at the Place Stanislaus, and found these students to be charming and very friendly. It turned out that it was very difficult to overwork one's self in a place like Nancy; even more so in the summertime.

CHAPTER IV

As Ramsay's Junior Collaborator

The Beginnings

In early August 1894, Morris returned from Nancy and joined his family at Lynton where they had rented a house for the month. Late in September he paid a visit to University College to re-introduce himself to Ramsay [whom he had written to asking for advice on his career]. All that Morris knew about the discovery of the new gas in the air by Ramsay and Lord Rayleigh was from the newspaper accounts of the announcement that these two men had made at the end of the meeting of the British Association at Oxford.

He found Ramsay working in a room in the basement called the "combustion room" where there was a slate bench on one side, with a hood over it, designed for making organic analyses. Ramsay remarked, "Well, it's a new element", and subsequently described the experiment very carefully.

Later, when leaving the laboratory, Morris asked Ramsay when did the laboratory open. Annoyed, Ramsay snapped back "On October 3rd. Good day". Travers felt that he had made a serious blunder here since Ramsay expected his staff to work at research in the laboratory for at least part of their vacation time. Morris immediately realized that he should have returned from Nancy earlier but he had no place of his own in the laboratory nor did he have any equipment. Furthermore, there was a previous strained history between the two men and Ramsay had only given Travers an appointment after long discussion with his wife [who was a close friend of Morris until her death years later] who virtually begged for Morris to be hired.

Morris's position turned out to somewhat unusual and difficult to understand. He was one of Ramsay's assistants and was paid, by Ramsay himself, an honorarium of £50 per year for lecturing to the junior students and demonstrating in the laboratory. In his spare time, Morris was expected to engage in research; independently or in association with Ramsay or a member of the staff; though at his own expense.

Fortunately, Travers's father was again quite generous to his son. William gave him a personal allowance and paid bills for books and equipment. Morris later found that, though there was an account set up in his name, his father regarded most of these payments as gifts.

With this funding, Morris started off by buying a Töpler pump, some mercury and glass tubing. He then took possession of two bench spaces in Ramsay's main laboratory and was assigned to "make" more helium by treating powdered clevite with hot, dilute sulfuric acid.

At about this point in time, Lord Rayleigh and Ramsay were at odds with the scientific public for their refusal to confirm the work on helium they reported on at the British Association meeting. Though the great James Dewar stated that his work showed the new gas did not exist, there was no doubt amongst Ramsay's staff and students that a great discovery had been made.

All doubts were laid to rest on January 31st, 1895 where Ramsay and Rayleigh discussed, in front of a huge crowd in a lecture theater, their paper on "Argon". This was according to Morris an event in the history of science that he described further in his famous biography of Ramsay.

Travers was still intent, at this time, on specializing in organic chemistry under the direction of J. Norman Collie. However, this was soon to change.

Investigation of Helium

On Friday, March 22nd, 1895, Travers once again went into Ramsay's private laboratory which he had not been in since his "meeting" with Ramsay in early fall 1894. He found Ramsay sitting on a stool operating a Töpler mercury pump, and beside him on another stool stood was a retort stand with a clamp holding a Plücker spectrum tube. He switched on the current from an induction coil, handed Travers a direct-vision spectroscope, and asked him what he made of the spectrum of the light he saw in the Plücker tube.

Morris said that it seemed to combine the spectra of all of the alkali metals. Ramsay said nothing but Travers later learned that the great man had already made up his mind that the bright yellow line in the spectrum was not that of the sodium D-lines but was a line belonging to the spectrum of the new gas.

At this time, Ramsay, whether by intent or not, had put in motion a series of events that was to lead to Travers's long association with him. This began with the need to further work out the properties of the new gas since Ramsay was soon to lecture to the French Chemical Society in Paris on argon. To an entreaty from Ramsay to do so did Travers readily agree.

On April 17th, 1895, Ramsay, Collie, Travers and A.M. Kellas (well known as an explorer who later died on an expedition to the mountain Kanchenjunga) met in Ramsay's laboratory. Ramsay, Collie and Travers began to work on helium while Kellas was involved with the measurement of the argon content of air and an attempt to detect argon in animal and vegetable matter. Morris felt uncertain of himself at this point; he had no experience in working with gases—what he had learned from working with acetylene was negligible.

Morris was to determine the density of helium by weighing the gas in a density bulb of 160 cc. Ramsay gave him the density bulb and the counterpoise bulb that he had used to determine the density of argon and a rough sketch of the apparatus that he had used in this analysis. When Morris got to his work place, he noticed that the counterpoise bulb was broken and its bulb was open to the air. To bring its volume and buoyancy to that of the density bulb, the counterpoise bulb had to be re-sealed which Morris proceeded to do.

The next day Ramsay came to see Travers and Morris told him what had occurred. Ramsay, silent for a moment, then said "How silly of me, of course it should be sealed, but I've used the bulb as you found it for all the work on argon". Ramsay could have ascribed this to breakage after the argon work had been finished but instead noted his noble character by stating that he had made a "silly" blunder.

This first series of investigations on helium covered the period from the middle of April till the beginning of June 1895. It consisted mainly of obtaining samples of gas from minerals, either by heating them alone, boiling them with dilute sulfuric acid or fusing them with sodium bisulfate. This work was directed at answering three questions:

1.   Was helium, which appeared to have a density a little over 2 (Oxygen 16) identical from all minerals and independent of method of preparation?
2.   Was the hydrogen content of the gas related to the helium content of the gas?
3.   Was the presence of helium in any mineral related to the nature of its elementary content?

Answers to these questions were never found but Morris and his co-workers spent a great of time in these researches. One fact that did emerge was that all the minerals yielding helium contained uranium with the exception of monazite, which happened to be a phosphate of cerium, lanthanum and thorium. The significance of this discovery emerged further in the researches of Rutherford and Soddy in 1900.

In June 1895, Collie went to India to attempt to climb, with two close friends, Nanga Parbat. There was however an unfortunate accident to one of his friends which delayed Collie's return to England well into November thus leaving the work to Ramsay, Travers and Kellas. With the return of Collie in November, the main interest of Ramsay became answering the question "Were argon and helium pure elementary substances?"

Morris had attempted to fractionate these two gases through absorption by platinum in a Plücker tube. This was not successful though. Knowing this, Ramsay [and Collie] tried to fractionate both gases by a diffusion process. Results from their work showed that the two gases were sensibly homogeneous and that argon had an atomic weight higher than had been anticipated from the Periodic Law and that both gases might be assumed to belong to the eighth group in the Periodic Table. It followed that, in all likelihood, there must also exist an element [as yet undiscovered] of atomic weight 20.

At this point, Morris, who still harbored the idea of going to Germany to study for his Ph.D., was invited by Ramsay to join in the search for this new element. Morris accepted this offer but later regretted the opportunity to learn German—which he found later in his career to be a significant handicap.

Ramsay and Collie continued to carry out their diffusion experiments while Morris was examining the gases given off by the heating of minerals, meteorites, etc. Morris found that more often than not these released gases were carbon monoxide and dioxide, water vapor and hydrogen. The carbon dioxide seemed to be present as carbonates, water vapor from hydrated silicates; though the carbon monoxide and hydrogen were not present as such, but were formed by the reduction of carbon dioxide and water vapor by ferrous oxide, present as silicates.

Morris also thought at one point that he had found a relationship between the quantities of helium and hydrogen given off by certain minerals. Though much effort was expended on studying this possible relationship, it all turned out for naught.

The spring term of 1896 had all very excited over the discovery of X-rays by Röentgen and the subsequent making of and experimenting with Crookes tubes by research students and staff alike. At Easter of this year, Ramsay and Travers made a trip to the Pyrenees to collect gas from the mineral springs at Cauterets that had been found to contain helium. The conclusion of this session saw J. Norman Collie become Professor of Chemistry at the Pharmaceutical Society's Institution and J. Wallace Walker join the staff.

Discovery of Krypton

During the 1896-1897 session, Ramsay and Travers were involved in attempting to fractionate approximately one liter of helium by diffusion. Their efforts though only obtained a trace amount of argon.

In September 1897 as President of the Chemistry Section of the British Association meeting at Toronto, Ramsay gave the address "An Undiscovered Gas". This address, predicting what would be arguably Ramsay's greatest scientific discovery, was also an outstanding piece of chemical literature. In it, Ramsay did not use the Periodic Law to argue that an element of atomic weight between the atomic weights of helium and argon existed. Instead, he favored the older and simpler generalization of Johann Döbereiner who thought that similar elements usually existed in groups of three where their combining weights possessed the numerical peculiarity that one was the mean of the other two. This, of course, allowed Ramsay to ignore the apparent abnormality in the atomic weight of argon.

At this time, Dr. Johnstone Stoney was working on the limits of the molecular weights of gases that could be retained in the atmosphere of the earth and other planets. It was suggested that Ramsay's "undiscovered gas" should also be present in the atmosphere of the earth. Morris thought that it should be present in samples of argon but in quantities too small to be detected by the methods used by the Ramsay group.

Consequently, a new approach was necessary. It was proposed to make a very large quantity of argon [15 liters], liquefy this "gas" and attempt to separate a possible low weight constituent by fractional evaporation of the liquid.

This proposal to separate argon into its possible constituents by liquefaction and fractional evaporation had been made possible by the work of the engineer William Hampson in England. Hampson, working concurrently with Karl Linde [in Germany], had devised a new method for liquefying gases.

In this method, a gas under high pressure passed through an approximately $1/8^{th}$ inch internal diameter closely wound coil of copper tubing that was enclosed in an insulated metal cylinder. The gas expanded freely from a valve at the lower end of the coil, and underwent cooling according to the Joule-Thompson effect. This expanded and cooled gas flowed upwards through the coil, cooling the downward flowing compressed gas, so that the cooling was progressive until the gas partially liquefied.

However, even with these developments, the 1897-1898 session was not moving forward to the satisfaction of Ramsay nor Travers. Their efforts in "making" the argon had taken longer than expected and they were almost ready for experimentation with this material, which they did not have!

Hampson then unexpectedly arrived in their laboratory on the morning on May 24th, 1898 with a cylindrical vessel containing about 1/2 liter of liquid air. Ramsay and Travers, though happy with Hampson's visit, had made no preparations for liquefying their argon. In fact, neither of them had ever handled liquid air and needed to get used to working with it. They then carried out a variety of simple experiments with it, including: freezing rubber and burning cotton wool that was saturated with liquid air.

Ramsay, as he often did, then had an inspiration!

Their liquid was to be allowed to evaporate until only about 7 cc of it remained. At this point, the Dewar vessel was closed off with a rubber stopper with tubes passing through it by means of which the gas from the least volatile fraction, which was mainly oxygen, could be collected. It did not seem that this was what Ramsay and Travers were looking for but the experiment was worth carrying out anyway.

The morning of May 31st began with passing this "new" gas over hot copper, and then over a hot magnesium-lime mixture, and then over copper oxide. The inactive residue, mixed with oxygen, was set to spark over the lunch hour.

Morris, who had lunch with J. Wallace Walker and had filled him in on what he was doing, was asked by him as they each returned to their laboratories after eating, "New gas this afternoon, Travers?" Travers, laughing, said "Sure thing".

The removal of the oxygen was carried out and a little of the "new" gas was introduced into a spectrum tube. Upon observation, the spectrum, though showing the characteristic argon lines, was dominated by a very bright green line and by a very bright yellow line with a greenish tinge. The tube obviously contained argon and also, with equal certainty, a hitherto unknown gas!

Ramsay, who had an unfinished letter to his wife (who was in Scotland) on his writing-table, added to the letter by stating: "Hurray! New gas out at last. Keep it dark". He then sent her a telegram, and later added to the letter:

> "The residue of 0.75 liters of liquid air, purified from oxygen and nitrogen, gives a line, nearly, but not identical with the helium and sodium lines in the yellow, and a green line identical with nothing else, besides others. I am wiring home that I won't be home to dinner, and Travers and I will do density and spectrum at once, and send a note to the French Academy on Friday. We have enough to characterize it, though sample contains a trace of argon. There is any quantity at our disposal, from the large quantity of argon we have. Perhaps the R.S. [Royal Society] on Thursday next might be the best means of publication. It might do good, not to go first to the Institute . . ."

Mrs. Ramsay's reply was quite interesting:

> "I would like to write you each [Ramsay and Travers] a letter, but time won't permit, I fear. Your telegram came in while we were at dinner last night. You get a new element every time I come away! . . ."

As May 31st, 1898 wore on, Ramsay and Travers worked with the 26 cc of the "new" gas that they had and found it to have a density of 22.47 while argon had a density of 19.95. The ratio of the specific heats was found to be 1.67.

It seemed certain that they had discovered a *new*, inactive constituent of the air, of density greater than that of argon. This new gas *was not though* the "undiscovered gas" noted by Ramsay in his previous address in Toronto.

When they finished their work on the new gas at about 11 P.M. that night, they both went home. Upon his arrival home, Travers was admonished *severely* by his father who told him that he had forgotten that he was to take a written examination the next afternoon for the D.Sc. degree at the London University.

On the morning of June 1st, effort was spent in writing a note for the Royal Society and sending a wire to Moissan for the French Academy, through which the discovery was made public before the weekend was finished. Ramsay, in a note to

Lord Rayleigh, said that he had decided on the new element's name as krypton; they had after all considered and rejected this same name for argon.

On June 2nd, their "krypton" was again sparked, and the density was found to be 22.5. A week later, their paper on this discovery was read to the Royal Society. Given that news of the discovery, emanating from Paris, had already appeared in the press enraged several fellows of the Royal Society. They stated that the Society had been insulted because the discovery had not been disseminated first through a letter to Nature. Consequently, at the reading of the paper, the new gas received hardly a word of welcome.

In addition, a series of letters began to appear over the signature "*Suum Cuique*" [To each his own] on June 17th in The Chemical News that first attacked the content of the paper read on June 9th and then attacked Ramsay himself. This series of attacks was annoying but nothing more.

Discovery of Neon

In the meantime, an experiment was conducted that gave a strong indication that air did contain an unknown gas lighter then nitrogen. Perhaps the "undiscovered gas" was soon to be found!

On the afternoon of June 7th, Hampson brought a second Dewar of liquid air to the lab of Ramsay and Travers. Serendipitously, he had been able to use one of the oxygen compressors at his works during the dinner hour. The large quantity of argon still present was not ready for treatment; but as there was a large gasholder full of atmospheric nitrogen and some of this was liquefied in a bulb cooled with the liquid air boiling under reduced pressure. The amount of nitrogen condensed was not recorded.

The spectrum [showing groups of lines in violet, red and green] of the most volatile portion gave a strong indication of the presence of some gas *other* than helium or argon. Given that Dewar was working toward the same goal of finding this "undiscovered gas", there was a concerted effort to discover this gas first.

The next step was to condense all the argon in a glass bulb cooled in liquid air boiling under reduced pressure, and to separate it from the most volatile and least volatile components. Though this procedure was new to Morris, it was easily mastered and a trial experiment was done. With the subsequent arrival of Hampson and more liquid air, the apparatus was tested and it turned out that the gas liquefied easily and did so quite quickly, at least to start with. After about three liters had condensed the current began to turn; and if the pressure on the liquid air was allowed to rise, gas began to boil back violently. Suspicion was the argon had frozen. Though this may have occurred it was possible to collect the "rapid boilings" in two separate 40 cc quantities.

This collected gas, fairly pure argon, was sparked with oxygen for 45 minutes. The spectrum showed red lines, a pair of green lines, a beautiful group of six lines in the blue, and a similar group of violet lines. The gas was then weighed and its density was found to be 17.24.

They were on the track of Ramsay's "undiscovered gas" but it was nearly midnight and time to go home.

On the afternoon of Saturday June 11th, Hampson came to the laboratory with another supply of liquid air. The apparatus was arranged as on the 10th and a total of seven different gas samples were gathered during the day and taken to Ramsay's room where they were set for further work on the 12th.

What occurred next depended on the spectroscopic analysis conducted by E.C.C Baly on the 12th. After preparing the needed equipment, analysis of the most volatile fraction produced a blaze of crimson light that was direct evidence of the long-searched for "undiscovered gas" of Ramsay! All the work had paid off!

After coming back down to earth again, Morris and his colleagues began to conduct work on the necessary physical measurements of the un-named "new" gas.

First, the "new" gas had to be sparked with oxygen, to remove traces of nitrogen, hydrogen or carbon compounds, which may have been present, though none of them showed in the spectrum. Oxygen itself was removed with phosphorus.

After this "purification" process, the gas was then weighed and the density was found to be initially 14.67. Further purification through fractional evaporation led to a density of 13.7. It here seemed abundantly clear that they had filled in the missing space in the periodic table between fluorine and sodium and in the vertical column between helium and argon!

A name for this "new" gas was suggested by Ramsay's son Willie [who had come to the laboratory to see how krypton was made]. He had come just as there were indications that the "undiscovered gas" had actually been discovered. Willie inquired "What are you going to call the new gas?—I should call it novum." His father said, "Neon would sound better".

Thus, the name for element with atomic number 10 and atomic mass 20 came about!

That same evening, Ramsay and Travers dined at the Ramsay home and after eating went to Ramsay's study to write up the paper for the Royal Society on the discovery of neon. It turned out that Ramsay's study, like his laboratory, was always in a state of disorder. Morris, having nowhere to sit, had to clear off a chair filled with papers. Lo and behold, the papers were the graded D.Sc. examinations! What a way to find out you had passed an important examination!

From this point on, further work was done on liquefied air, determination of the properties of argon, krypton and neon and separation of the argon/krypton and argon/neon mixtures. It turned out that pure neon could not be obtained through use of liquid air; it was thought that liquid hydrogen would be needed to accomplish this separation.

Discovery of Xenon

These efforts all led up to the evening of July 12th where Morris and others had been working at the fractionation of some argon-krypton residues when, after removing the vacuum vessel from the liquefying apparatus, which had been pumped out, a bubble of gas [probably CO_2 in Morris's opinion] was seen still in the pump.

It was late and there was just barely enough time to catch the last train home. Ramsay was standing in the doorway with his hat on when Travers had the idea to remove the liquid air, and collect any small quantity of gas that which remained in the apparatus. Ramsay went to the train and home; Travers continued on and collected the gas, missed the last train and had to walk home to Kensington.

Morris continued on July 13th by examining the fractions of gas obtained from the previous evening's work. The gas in all the fractions gathered seemed the same spectroscopically. Fraction 56 showed very strongly the presence of krypton while fraction 57, chiefly CO_2, was obtained as the last fraction of a large quantity of argon.

After treatment with potassium hydroxide, the residue of about 0.3 cc was introduced into a spectrum tube. Krypton yellow was very faint while krypton green almost absent. Instead, several red lines, three brilliant and equidistant, and several blue lines were seen. Morris wondered:

> "Is this pure krypton at a pressure which does not bring out the yellow and green, or a new gas? Probably the latter."

The spectral tube gave out a brilliant blue glow. A name was sought for the gas based on Greek or Latin roots so noting this glow; however the ground had been worked over too well by organic chemists. The name xenon, the strange one, was selected for it.

Ramsay, who had been away from the laboratory for three days at hearings of the Royal Commission on the disposal of sewage, was actually *disappointed* rather than pleased with this discovery. This may have been due to the fact that the discovery was due to persistence and hard work on the part of Travers rather than a brilliant piece of scientific work. Or it may have been more likely due to the fact that the discovery of xenon had extinguished Ramsay's hope that the rare gases could be shown to exist as triplets and thus confirm Dobereiner's ideas.

CHAPTER V

Liquid and Solid Hydrogen

Vacation and Tragedy

The summer vacation of 1898 was one that had a profound impact on Morris's life. During this time, Morris had been walking and scrambling rather than doing the serious Alpine climbing in Switzerland he had intended to with his brothers Ernest and Cecil.

In early September, Ernest and Morris were at Arolla. Fellow guests at their hotel included Professor John Hopkinson (an electrical engineering Professor at King's College in London) and his family [even including his son's tutor]. The tutor, Ernest and Morris were to climb the Aguille de la Za with Morris leading. Professor Hopkinson, his son and two daughters set off at the same time to climb the Petit Dent de Vesivi. The two parties left each other at about 4 A.M. for their respective climbs and jokingly planned to meet later at tea time.

Morris thought nothing more of them and had forgotten about this "planned" meeting. Playing whist after dinner Morris was asked by Gilbert Walker, to come outside where Walker whispered "The Hopkinson party aren't back yet". Morris and Gilbert saw no lights on the mountain and returned inside to find an extremely distressed Mrs. Hopkinson begging that a search party be formed immediately.

Morris, who knew the Petit Dent de Vesivi very well, and three tour guides formed a search party and began a search for the Hopkinsons. It was Morris's intention to leave Arolla early the next morning. He had packed all of his clothes except for what he was wearing and the sweat-laden clothing that he had worn on his day's climb. In his haste, he made a very poor decision and put on the sweat-drenched items.

The search party immediately began its efforts but when they reached the rough ground, over which Morris could not lead them at night, they stopped and rested until dawn. This effort thoroughly fatigued Morris and gave him a bitter chill that persisted for a good while. In his opinion, this effort was instrumental in the serious illness he was soon to experience.

Unfortunately, their search efforts only led to the discovery of the Hopkinson bodies. After finding them early in the day, they then returned to Evolena. Later in that same day Morris, being the only Englishman who spoke intelligent French, had to go and meet the local coroner. This meeting did not occur until late that day and did not help Morris's overall health.

Back at the University

In September 1898, Ramsay and Travers described the discovery of neon, krypton and xenon to a meeting of the British Association for the Advancement of Science [now known as the British Science Association] in Bristol. The first two terms of the 1898-1899 academic year still had Morris and his colleagues engaging in most of their work from 7-12 P.M. but the gift of a Hampson air liquefier [from Ramsay's friend Rose-Innes] and of an air compressor and motor by Ludwig Mond enabled them to now finally obtain liquid air at any time. This made everyone's life much simpler.

They were able, during these sessions, to obtain a sufficient amount of reasonably pure krypton and xenon to find their densities, fix their atomic weights, and show that they occupied their predicted positions in the Periodic Table. In addition, by liquefying air, and subjecting the liquid to fractional evaporation, they obtained more of the mixture of gas rich in helium and neon without repeating the labor-intensive task of "making" more atmospheric argon.

Morris had no doubt that neon was the gas of atomic weight 20 which they were searching for. But he and his colleagues could see no way to separate it out pure.

The beginning of the 1899-1900 school year saw further attempts to separate out the neon without any success. By the end of 1899 it was clear that the only solution for Morris's problem lay in using liquid hydrogen to try and make this separation.

Liquid hydrogen

At a meeting of the Chemical Society held on May 10th, 1898, James Dewar said he had obtained liquid hydrogen in a static condition. In his submitted paper though, he only described an apparatus through which he had, at some earlier time, obtained a spray of the liquid. Olszewski, a Polish chemist working with Wroblewski, also had made a similar claim in 1895. Dewar said that all attempts to collect the liquid into a vacuum vessel failed but the apparatus worked well enough so that he resolved to build a much larger liquid air plant and combine it with circuits and arrangements for the liquefaction of hydrogen. This apparatus took about a year to make and many months were needed to test it and conduct preliminary trials with it.

This was clearly the apparatus used to obtain a small quantity of liquid hydrogen as he noted on May 10th. However, no description was given of the apparatus at that time nor was it given later. An account of Dewar's success in liquefying hydrogen and helium was communicated to the Daily Chronicle. Lord Rayleigh was also shown the colorless liquid but he told Ramsay that there was no evidence what it was.

Given this, Morris concluded that he was not undertaking an easy task in attempting to liquefy hydrogen but he felt this to be a condition that favored the adventurous.

From when they decided that they needed liquid hydrogen, Ramsay, who disliked anything connected with machines, left the work to Morris. Morris, who delighted in "playing" with mechanical toys when he was a child, had acquired a talent for soldering that led to his pipe joints never leaking and he was certainly up to this difficult task.

Morris began this work by reading the literature that commented on the cost, difficulty and danger in attempting to liquefy hydrogen. He began his work by buying a gas storage device made from steel sheets and angles, riveted together with the joints covered with tar, which had a 120 cubic foot capacity. This cost £14. Unfortunately, the tar never set hard and the joints always leaked. This was mistake number one and it was due to Morris's inexperience. Morris then bought a beer cask for collecting hydrogen gas from the reaction of zinc metal and sulfuric acid. He did not realize hydrogen would very rapidly diffuse through wood; this was his mistake number two. Several coats of paint on the barrel dealt with this problem. But he had not reckoned with the fact that if any hydrogen left the barrel by diffusion; some air would enter in its place. This was his mistake number three.

Having air in the hydrogen nearly caused the failure of Morris's first two attempts to liquefy hydrogen. Before his third experiment a year later, Morris bought a lead vessel for a hydrogen generator and he had never had trouble with air in his hydrogen again.

An obvious source of danger was in the compressor, which was a two-stage machine with a single acting low-pressure cylinder with a cup and leather packing. Air might be readily drawn into the cylinder around the packing material. To correct this, the cylinder was lengthened, and a piston with rings was added behind the piston with the cup packing—with the space between the two pistons being filled with water. The heavy work needed to put all this together [making the connections with steel pipe between the gas holder, the hydrogen generator and the compressor] was carried out by a friend of Morris's who served normally as the lecturer's assistant.

In Morris's first experiments to liquefy hydrogen, he used an old Hampson liquefier, to which he had added a coil of copper tubing, which was immersed in liquid air, and through which the compressed hydrogen passed before entering the liquefier. The regenerator coil of the liquefier had also to be cooled with liquid air before starting an experiment.

Morris's friends in the laboratory predicted that he would blow himself up, a contingency that he considered at least remotely possible! Interestingly enough,

Morris found no one wanted to be with him in the compressor room when he conducted his first trial.

This initial effort to liquefy hydrogen had very little chance of success but after carrying it out Morris and his colleagues felt confident that working with hydrogen was neither difficult nor dangerous.

It was thus apparent to all that Morris would have to design and make a liquefier. Having an engineering firm do so would be too costly. Morris had never made an engineering drawing in his life, though had a very clear mental picture of what he wanted. Details could be worked out along the way.

Morris wanted an exhaust pump and some means to drive it. A student assistant had made a blowing pump in his workshop for producing compressed air for use in glass blowing. He offered to loan this pump to Morris and suggested that Morris reverse the valves and use it as a vacuum pump, which he did. The Engineering Department of the college also helped by offering the loan of an early form of a gas engine as a source of power; Morris was able to make this "broken" unit work and use it for his experiments in July 1900. This unit was later replaced by an electric motor.

Getting this work done quickly was an overriding priority for several reasons. First, Morris was becoming increasingly ill, with a gastric ulcer, originating from the Alpine accident that claimed members of the Hopkinson family and too much work in the many months following.

Second, in his position as an assistant examiner in chemistry for the London University, Morris had to grade some two thousand answers to questions from the elementary general science papers by candidates for the London matriculation. Though this grading could be done on top of the mechanical work on the hydrogen liquefier, the end of the first week in July would see the opening of the intermediate examinations, which would be even more time-consuming in their marking. This marking was essential for Morris and others; they made half of their incomes by examining students at the university. They could not afford to give this up!

Third, Ramsay had informed everyone that he would be going to India in the fall and would be gone for some time.

And so came Saturday, June 30th, 1900 with the apparatus ready for a trial. The gasholder was filled with hydrogen and the plant was started up. All went well at first. The pressure was raised to about 180 atmospheres, and the expansion valve was opened. Soon a hurricane of snow-like material was seen in the bottom of the vacuum vessel. The came out a small amount of liquid which was evidently hydrogen, but this was only a small amount. At this critical moment the compressor broke down and ended the experiment.

One week later on July 7th, they were ready to try again after a careful overhaul. With a fresh supply of hydrogen and a new supply of liquid air, all went well at the start of the experiment. The valve was opened, the hydrogen was permitted to expand and a snowstorm of solid air and other impurities from the hydrogen were seen in the vacuum vessel. Then liquid began to drip from the valve. It was though only a trickle and soon the valve became blocked. Nothing further was coming out of the valve and the pressure was quickly increasing.

Morris now had to think fast. Should he hold on and hope to blow out what was blocking the valve or stop the experiment, allow the apparatus to warm up and so free the valve? Morris, having only a small amount of liquid air on hand, decided on the former course. The pressure rose to over 250 atmospheres with Ramsay screaming in Morris's ear to stop before disaster would occur! The mere fact that Ramsay was standing there probably was reason enough for Morris to hold on and to hopefully blow out whatever was blocking the valve.

Just at the point where Morris would go no further did something happen with the valve. A loud hiss indicated that hydrogen was expanding again. Liquid hydrogen was filling the two vacuum vessels! All that remained was to stop the compressor, detach the receiving vessel, plug it with wool, and put it inside a larger vessel containing liquid air.

Though this work had taken more than a month to complete, it was only a means to the end of using this liquid hydrogen in a different set of experiments; the purification of neon.

Neon

The hydrogen was immediately carried up to Ramsay's room, where, in anticipation of successfully obtaining the liquid, all was prepared for the final experiment. Though there were a variety of fractions separated out in this final experiment, the objective was to obtain a small sample of neon free from any trace of helium, which would affect the density significantly. The fact that argon was for all practical purposes non-volatile at liquid hydrogen temperatures made it obvious that the separation of neon and argon did not require any special conditions.

This was the last experiment that Morris ever carried out with Ramsay. In it, the Töpler pump was set up on the end of the table and Ramsay manipulated it and collected samples of the gas. Travers attended to the liquid hydrogen and the operation of the fractionating apparatus. Little was said until the tube containing 15 cc of neon was standing safely in the rack.

Though they had obtained liquid hydrogen and pure neon for the first time in one day, extreme exhaustion ruled the moment for Morris and Ramsay. No notes were made and the hydrogen liquefier and fractionating apparatus were left as they were. Notebook entries, made on Monday the 9th, were very brief:

> "Saturday, July 7th, 1900.
> "Fractionated Ne He in liquid hydrogen. (About 75 c.cs. of liquid hydrogen obtained).
> "Fractions of Ne He mixed. Liquefied easily.

"First fraction through pump, 1/2.

"Second fraction . . .

"Third fraction through pump after removal of hydrogen.

"Second fraction introduced into gas-holder and reliquefied—vapour pressure only a few mm.

"First fraction (about 1/10) separated.

"Second fraction (main quantity) probably neon, taken through pump.

"Residue through pump after removal of hydrogen.

On July 9th and 10th, Morris and Ramsay each made an independent determination of the density of the collected gas fraction. Morris found it to be 9.94 while Ramsay obtained 9.99. Ramsay, then from the mean result of their two experiments, wrote down a value of 19.975 for the value of the atomic weight of neon and added Q.E.D. and said to Morris, "Travers, this will not be repeated for twenty years." This statement proved to be prophetic.

While engaged in this last bit of work on neon, Ramsay told Morris why he was going to India. Ramsay was to examine a proposal by the wealthy J.N. Tata to found an institution for scientific and industrial research. Ramsay said to Travers: "If the proposal materializes, would you like to go out to India as first Director?" Morris refused to consider the idea.

Illness and First Book

While Ramsay completed determination of the compressibilities of the gases, Morris tried to return to the marking of London University examination papers but had to give it up due to his stomach problem. His good friend Donnan took him away to his home near Belfast where he and his family started him on the way back to health. Sir William Osler, the esteemed Canadian physician who is often regarded as the father of modern medicine, noted to Morris in Bristol in 1905:

> "I consider it a great advantage to a young man to have a serious illness early in his life. He realises his limitations."

Morris took this to heart and used this advice later in his life when facing other medical problems in 1912-1914, 1916 and in 1955.

When the fall 1900 term began, Morris acted as Professor while Ramsay was in India. He was fortunate that Baly and Donnan took most of the extra work off his shoulders. In fact, Morris did not feel well enough to begin research again until after the 1901 summer vacation.

During the last year's work on the rare gases, Morris, in addition to engaging in research and examining, was also writing a book. His first book, *The Experimental*

Study of Gases, was well received and is regarded as a chemical classic in 2011. Morris would rewrite it for translation into German by Estreich. It was published in German with the title *Experimantelle Untersuchen von Gasen* in 1905.

Low Temperature Work—Continued

Morris decided that when he returned to experimental work that it would be to exploit the field, which his success in liquefying hydrogen had opened up. He though first had to make some minor modifications to the apparatus he had built.

He felt the next thing to do was to establish a scale of temperature, and he proposed to do this by measuring the temperatures corresponding to the vapor pressures of liquid hydrogen and liquid oxygen on the constant volume hydrogen and helium scales. Morris even planned, in collaboration with Lehfeld, to conduct a general study of the thermodynamic properties of hydrogen and helium [this never did occur though].

Morris made a very careful examination of the experimental data found in the most recent investigations using the constant volume system and discovered that a most important source of error lay in the control of the temperature of the mercury column of the measuring instrument. He decided to eliminate this by enclosing the column in a water jacket where the front and back were plate glass and the scale was on the front. He felt that the apparatus was rather crude, given that part of it was made of wood and that it needed adjustment to the vertical by means of laths fixed to the wall of the room. This wood structure allowed Morris to obtain accurate results, but at the expense of much wasted time from making adjustments.

It was Morris's belief that this was the first work in which gas pressures by a mercury column were ever accurately measured. Results of this work were communicated to the Royal Society in June 1902 while a paper on this work appeared previously in the Philosophical Transactions in 1901. These efforts noted the boiling points under standard atmospheric pressure, on the constant volume hydrogen and helium scales, in degrees Celsius for both oxygen and hydrogen. In addition, the pressure coefficients of both hydrogen and helium were found to be 0.0036625, which is the reciprocal of 273.05 [the value of the ice point in Celsius on the two scales].

Morris's collaborators in these low temperature measurements were George Senter, later Principal and Professor of Chemistry at Birkbeck College and Adrien Jaquerod who continued to work on thermometry.

Making liquid hydrogen and lectures

During 1902 and 1903 the making of liquid hydrogen became a routine matter that always succeeded. It always attracted an audience including Ramsay, if he were in at the college. Guests were frequent and always welcome and one of Morris's student assistants was always ready to answer any questions about the process. This was consistent with the fact that there were no secrets in the ever-growing Ramsay "group"; Morris was also anxious to bring others into low temperature work and to initiate workers from abroad in this field.

Morris, acting on an invitation from the University of Bonn [where the German Crown Prince and his brother were studying], designed and installed in the Chemical Institute a plant for producing liquid air and liquid hydrogen. In addition to this, Morris gave two lecture demonstrations for the benefit of the Imperial Prince and his Court in January 1903. He showed them liquid hydrogen burning at the top of a glass cup, suspended so that solid air formed on the bottom of it. Morris, who had also taken with him a constant-volume thermometer, also, for them, measured the temperatures of ice, dry ice, of liquid air and of liquid hydrogen.

May 1903 had Morris, during the meeting of the Congress of Applied Chemists, make liquid hydrogen in the Technische Hochschule at Charlottenburg in Berlin. To do this he used his own liquefier operated with hydrogen gas compressed into cylinders at 125 atmospheres. It was the first time Morris had liquefied hydrogen in this way.

In November 1903, Morris gave two evening lectures at University College. The first demonstrated, through experimentation, the preparation of the five rare gases (helium, neon, argon, krypton and xenon). The second dealt with his work at low temperatures and showed the audience liquid and solid hydrogen. Liquid hydrogen was contained in a triple-walled but un-silvered vacuum vessel. It was then stoppered and connected by thirty feet of rubber tubing to a vacuum pump. The liquid froze when the pump was turned on. The solid hydrogen had the appearance of snow that had partially evaporated without melting and was free enough in the vessel that members of the audience handled it without leaving their seats.

Robert Whytlaw Gray

A short time after these lectures, Robert Whytlaw Gray began to work for Morris as a research student. Gray had previously entered the department in 1898 but had been badly injured in November of that year while attempting to prepare saccaharin through a diazotization reaction that exploded in his face. Morris was the one who took Gray to the hospital for treatment and then later visited his family's home to warn them that Gray would probably suffer from shock though his injuries had been treated. It was here that Morris first met Gray's sister Dorothy [then 14] who would become Morris's wife in 1909.

At that time, Gray had to give up the idea of a University College London degree. He only came intermittently to the laboratory until 1901. He then gave an excellent paper on experimental work on gas density determination to the College Chemical Society.

Morris then thought that Gray might apply the technique that he was using in gas thermometry to the accurate measurement of gas densities at varying pressure, with the object of determining the values at infinite expansion, and from them the values of the true molecular weights, and in the case of elementary gases the atomic weights. Gray took up this work and found, in his first investigation, that the atomic weight of nitrogen was 14.01 and not the generally accepted 14.03.

Gray then went to Bonn to study with the great German chemist Richard Anschutz and received his PhD in 1906. Gray then returned to University College London where he worked with Ramsay and Soddy in the characterization of radon. He later taught at Eton and became Professor of Inorganic Chemistry at Leeds University.

"Leaving the Nest"

Financially, Morris was earning close to £175 as an Assistant Professor with approximately another £100 as an examiner at the London University while his father generously paid for Morris's work related expenses.

Until the autumn of 1901, Morris had been happily living with his mother and father in Kensington but his brother Harold came home from South Africa a very sick man battling typhoid fever and needed additional space and attention. So, Morris left home. He first took rooms in a house on Finchley Road and then in the spring of 1902, he moved into a place facing the north side of Regent's Park. This place, a rather ruinous first floor, was cheap. Interestingly enough, his window remained propped open with an ice axe during his time there! Morris's life at this point was still restricted to his work.

In the summer of 1902, Morris went on a summer vacation with one of his brothers in Dauphine [southeastern France]. Upon his return, Morris took a small flat in a street behind the east side of Portland Place. This cost him £50 a year as well as a bit more he paid to a woman for cleaning, laundry, etc. Morris made his own breakfast, got lunch down near the college and cooked his own dinner. He still had lunch and dinner on Sundays at his father's house, which he considered "home" until Christmas 1904.

CHAPTER VI

Professor of Chemistry and the University Question

Seeking a position

From 1900 on Morris had begun looking for an independent appointment while continuing to work with Ramsay. He applied for and was rejected for the Chair at the Pharmaceutical Society's Institution in Bloomsbury, and for the Lees Readership in Christchurch, Oxford. Neither of these rejections bothered Morris much though.

In November 1903, Sydney Young, who had succeeded Ramsay in the position of chair of chemistry in University College Bristol in 1887, was selected for the chair of chemistry in Dublin. Morris, when this position in Bristol was advertised, was not much interested in it.

His lack of interest was understandable for many reasons. Though he had spent a happy week in Clifton during a meeting of the British Association for the Advancement of Science in September 1898 and knew that this was a pleasant place to live, and that the people of Clifton regarded the college staff highly, Morris had no friends in Bristol and had made no further contacts even with the Tyrons [old friends of Ramsay], who had been his hosts in 1898. Morris was also a Londoner through and through; and disliked the idea of being banished to the provinces.

Morris was also well enough acquainted with university education in Great Britain to know that while Bristol University College had an academic record equal to that of the colleges in Manchester, Liverpool, Leeds, Birmingham and Sheffield, it had not been well supported financially from its beginning. Bristol received less annually from the Chancellor of the Exchequer and had the lowest income and endowment out of all the noted institutions. In fact, the salary offered for the appointment was only £350 per year, which was a low salary for a professorship.

Morris asked Ramsay for advice. Ramsay, agreeing with Travers for the most part, noted though that it would be an independent command; and if Morris did well in it that he should not remain long in Bristol unless he wanted to. Ramsay also said he had been assured by friends in Bristol that significant progress had been made on a plan to obtain a university charter for Bristol and that support for this endeavor was widespread.

Morris, somewhat reluctantly, applied for this position at Bristol. His application was supported by reference letters from a number of British and foreign sources though this carried little weight with the selection committee. The committee though put Morris on the "short list" and he was invited to meet the council of the college on November 18th at mid-day.

He traveled to Bristol by evening train. Morris arrived at his hotel at about 10 P.M. and learned that it was closed! He was fortunate though that he did get into his room that night. The following morning he asked for directions to the University College but was sent to the grammar school—apparently a better-known institution! Morris wandered about Clifton and a short time before mid-day presented himself at the college where the six candidates for the position gathered.

All the professors and independent lecturers from the Joint Faculties of Arts, Science and Medicine under the direction of its Chair Professor Conwy Lloyd Morgan (F.R.S.) interviewed the candidates. Morris was asked questions about his previous teaching and research experience. Then, a remark from the Engineering Professor R.M. Ferrier [later to become a close friend of Morris's], and the subsequent reply from Morris brought about an argument where Morris was alone in his views. Professor Lloyd Morgan ended this discussion; Morris felt amused at what had transpired. One of the Joint Faculties made a remark, which Morris later learned, that at least one of the candidates had views of his own and was not afraid to express them. When all six candidates had been interviewed, they were asked to return to meet the council again at 2 P.M.

When reassembled at 2 P.M. in a waiting room, the registrar informed them that only two names were being considered, Travers and E.F. Francis who was currently a Lecturer in Chemistry in the college. Francis, who was a very popular candidate, was first to meet the council. In due course Morris was summoned to meet the council, and was duly informed that he been chosen by them. Morris was asked if he accepted the position and he answered yes. He then bowed and left the room.

Morris heard afterwards that the Joint Faculty meeting had sent both the names of Travers and Francis forward to the council for a decision to be made. With the final decision up in the air, the Bishop of Bristol noted that Morris's London University doctorate had more value than the Victoria University doctorate held by Francis. This apparently settled matters; Morris won by one vote.

After leaving the council room with the position in hand, Lloyd Morgan congratulated Morris. Travers then went up to the laboratory to see Professor Sydney Young who was in his private room with Francis. Young was formal but pleasant; but Morris recalled Francis with tears of bitter disappointment in his eyes stepping forward with a smile of welcome and a hearty handshake. Till Francis's death in

1941, Morris reckoned him as one of his closest friends and a stellar example of what an English gentleman was. Francis would be Morris's best man at his wedding in 1909 and was godfather to his son Robert in 1913.

After sending wires to his father and to Ramsay, Morris returned to London by afternoon train. His appointment appeared in print and he received many letters of congratulations. Two were from Donnan and Robert Whytlaw-Gray, Morris's future brother-in-law. Donnan hoped that in due time Morris might return to his native city. Whytlaw-Gray, in addition to offering his congratulations and best wishes, commented whimsically on the abundance of religious holidays in Bonn that shut down the university and his work and on the fact that the preferred field of chemistry in Germany was organic chemistry and that his work barely merited curiosity.

On the day following the council meeting, Lloyd Morgan wrote Morris a long letter, probably as instructed by the council. He said that the council felt that Morris's ability as a research scientist had been stressed in his recommendation letters. He had assured them that Morris "would devote himself heartily to teaching, and keep close in touch with the laboratory progress of the students". Lloyd Morgan continued,

> . . ."As you may suppose, I am not one to undervalue research—nay, I hold that unless a professor is extending the boundaries of our knowledge, he will not win the confidence and admiration of his best students, and of his colleagues".

It was a kindly letter, and Morris could take no exception to anything in it; though he did feel that it would have been more highly appropriate had he been ten years younger. The letter ended with the words

> . . ."I don't think that Young and I have had the smallest misunderstanding during the 16 years he has been Professor of Chemistry. Absit omen".

Morris and Lloyd Morgan had no misunderstandings; Travers made it perfectly clear from the beginning that he had definite views as to what the policy of the college should be on the important question of its development into a university. They differed strongly; but remained good friends.

Conwy Lloyd Morgan, F.R.S.

Conwy Lloyd Morgan was born in 1852. He joined the college in Bristol in 1883 as lecturer in geology and zoology, becoming Dean when Ramsay was selected as Chair of Chemistry in University College, London in 1887. He had been granted this honor ahead of several more senior professors. Though he did some excellent work as a field geologist, he was best known for his studies in animal behavior, which put him at the forefront in the field of experimental psychology. In 1910 he was named the first Vice-Chancellor under the university charter; but he soon retired from this post to hold a chair in psychology and ethics in the university. He died in March 1936.

Morris regarded Lloyd Morgan as a cultured man with wide interests and a good sportsman. In fact, Lloyd Morgan was an excellent golfer, though he only took the game up later in life.

Though an ideal head of a college, founded to bring culture and learning within the scope of those who could not afford residence in one of the old universities, he was not a good administrator. He was though, as the social head of the college, popular with its staff and most of Clifton's society.

Of further note, in January 1906 in a social event at Lloyd Morgan's home, after a lecture on economic geography by Halford Mackinder who was a former Principal at University College Reading, the relative positions of the Reading and Bristol Colleges came up for discussion. Reading College, founded in 1892, had an endowment of £50,000 and an income of £15,000. The endowment at Bristol was £5,000 and the income was £10,000 though the college had been open since 1878 and Bristol was by far a wealthier city than Reading. Mackinder remarked to Morris that while at Reading as Principal, he had been an autocrat. Autocracy may have been a virtue in Mackinder; it would have been a vice in Lloyd Morgan, and autocrat he was not.

Position in Liverpool?

On a morning in November 1903 only a few days after his appointment at Bristol had been made, Ramsay came into Morris's private laboratory at University College London and told him that E.K. Muspratt, the well-known chemical manufacturer and Chancellor of the University of Liverpool was in Ramsay's room and wanted to see Morris.

Muspratt told him that he was proposing to build a laboratory and to endow a chair of physical chemistry in the University of Liverpool, and that he wished Morris to be the first professor in this position. The two men discussed this a while and Muspratt noted he had spoken to the scientific staff [including pathology Professor Rupert Boyce and physiology Professor Charles Sherrington—who would be the 1932 Nobel Prize winner in Physiology /Medicine] and that Morris would be well supported by them.

Morris, acting on an invitation by Boyce, visited Liverpool. He found many supporters on the staff that thought highly of his work at low temperatures and on the rare gases. Travers was much taken with Liverpool and the university was a live and active entity that was well supported locally and growing rapidly. There was also a very close cooperation between members of the university council and of the staff in matters relating to the development of the university, a condition very different from which, as he was shortly to learn, existed in Bristol. All in all, the atmosphere was totally different than what Morris had seen in Bristol.

However, there was a problem. It seemed that, according to the rules of the university, the position must be publicly advertised and that Morris had to formally apply for it. He told his friends that if he were offered the position, he no doubt could obtain a release from Bristol or even resign after three months though that would harm Bristol. Morris, being a man of honor, would not do that nor could he not let Lloyd Morgan know of his moral dilemma.

The appointment in Liverpool was a very attractive one with a salary of £500 per year and the possibility of making as much through consulting work. Every condition, save the university's need for a formal application, favored Liverpool.

This affair dragged on for weeks well into 1904 and was not resolved until after Morris was in Bristol. When he was setting out for an Easter holiday, Morris received a telegram from Boyce who asked if Morris would allow Boyce to put in a last-minute application for him. Travers stopped at a post office along his way and sent a reply to Boyce saying no. That finished his chance at a Liverpool professorship.

Travers later paid a visit to Liverpool to show them that he had no hard feelings over what had occurred and was asked by Boyce who he could recommend for this position. Morris mentioned F.G. Donnan of Dublin immediately and said to Boyce that Donnan was outstanding and that Liverpool should have realized this without Morris's recommendation [Donnan was later selected for this position and served as the Chair of Physical Chemistry until 1913].

When all of this was over, Morris told this story to Francis, remarking that he had not spoken of this matter lest it raise false hopes for Francis and the position at Bristol. In reply Francis stated, "I knew all about it, but I never gave it a thought, for I knew that you wouldn't apply for the job".

Leaving London/ Arriving Bristol

Students and staff from the Ramsay laboratory gave Morris a farewell dinner and gave him a going away present of some pieces of plate. Though these arrangements were on short notice, Morris was happy that a group of students, who had not heard about his departure, gave him a circular letter.

Morris spent Christmas day of 1903 with his mother, father and some of his siblings at Number 2 Phillimore Gardens, returning to his chambers late, and leaving the next morning to join a group from University College for a week's vacation in Cornwall. He had arranged for a firm to pack up and store his belongings, pending instructions for their delivery to Bristol.

Lady Ramsay had written to a number of her friends asking them to take notice of Morris. These were people of her age and standing, and were very kind to Morris, after all the formalities, usual at that time, had been gone through. Morris also had had a private tea with Lady Ramsay so that he could receive advice regarding Bristol, or more correctly Clifton. Morris learned that he must live in Clifton, for there he would find most of his friends and that the residents of Clifton had no dealings with the residents across the Whiteladies Road. She also stressed the good qualities of those people Morris should like rather than warning Morris of others. Last of all, with a twinkle in her eye she said to Morris, "You may possibly marry, but it mustn't be a Miss X or a Miss Y"—"Why, indeed," asked Morris—"Aren't they respectable?" "No,"—she said, "quite on the contrary",—and left him to find out the answer for himself, which he did in due course. He did not marry for another five plus years and then to a London girl.

Morris then looked around Clifton for a place to live and found convenient unfurnished quarters on the 1st floor of 8 Savile Place. After his furniture arrived, he was able to get his books and pictures in place before the session opened on January 16th. He hired the building caretaker Mrs. Arnold, who lived in the basement, to cook and clean for him. Several months later, he took over the whole house and shared it with his colleague Ferrier. The faithful Mrs. Arnold looked after the food supply and its preparation. Her faithful, but rather futile, husband cleaned boots and carried coal. This was Morris's very happy home until he left Bristol in 1906.

Chemistry at Bristol

Morris began by clearing up the debris that every new professor of chemistry finds that his predecessor has left him. Though Morris had paid one fleeting visit to Bristol after he was appointed, he saw nothing of either E. F. Francis or C.C.M. Davis. It seemed at first doubtful that Francis would stay to work under a younger man (Morris was not yet 32 years old). Morris was glad to learn on his arrival that the council had given Francis the rank of Assistant Professor and that he would be around at the opening of the January session.

Davis, the demonstrator, was a remarkable man in that he had no ambition for himself but had the keenest sense of duty and consideration for others. Morris was able to leave the junior students entirely to him, merely making suggestions as to methods of teaching, particularly in practical classes. The students started on quantitative, rather than qualitative work, carrying out simple experiments such as showed loss or gain of weight during chemical change, or the volumes of gases generated in chemical processes.

Davis was engaged in research work with Francis that ultimately led to his receiving a D.Sc. degree from the University of London. During the First World War, Davis became interested in medical problems and decided to qualify in medicine, taking an M.D. degree. But that was not all for this remarkable man! He developed an interest in forensic medicine, decided to read law and was later called to the

Bar. In medicine he specialized in pediatrics and became attached to the Bristol Children's Hospital. In fact, when Morris's grandson Christopher became critically ill soon after being born it was Davis, after others had bungled Christopher's care, who took personal charge and with two first rate nurses pulled the child through. Davis remained as lecturer in both chemistry and forensic medicine in the University until his retirement at the age of sixty-five.

Francis came to Morris on the first day of the session and asked him to lunch. At lunch they discussed the allocation of the work with Francis taking the share of it that would have gone to a Professor of Organic Chemistry, had he had that title. Morris immediately saw that he could treat Francis as a colleague of equal academic standing yet remain head of the department.

Evidence of this was seen a few days later when there was a meeting of the Faculty of Arts and Sciences. Morris asked Francis to come with him to this meeting! This was unheard of—having an assistant professor attend. Though the Principal raised an eyebrow, he must have guessed it was based on Morris's initiative; and the Principal had learned, and was to learn still further, that Morris had a mind of his own, which was an unusual quality for a professor.

Question of the University Movement

Early in the term Morris visited Registrar James Rafter and asked him about the university movement, of which Ramsay had spoken. He expressed total ignorance of the matter and said there were no papers about it in his office. He did though note that a Mr. Haldane [afterwards Lord Haldane] had made a speech on the subject as a guest at the annual dinner of the Bristol University College Colston Society [a charitable organization looking to raise money for the University College] in February of 1902. In this speech Haldane had advocated the formation of a university along the lines of Victoria University, which had included the University Colleges of Manchester, Liverpool and Leeds. In his model, the University at Bristol would include colleges at Bristol, Reading and Southampton. Morris soon got a copy of this speech which had made little impression on the Bristol public at both the time of the speech and afterwards.

Morris left the registrar's office both shocked and disappointed. He though returned a day or two later, after Rafter had told him that a report of the council and the Combined Faculties of the College to the City Council existed, but he could not show it to Morris without the permission of the Principal. Permission was given but no copy of the report could be found. Much later, Morris received a copy of the report from a colleague but he found that the "report" was nothing but a collection of generalities, and did not even contain the beginnings of a university scheme.

Rafter soon spoke to Morris of a committee that the council had appointed three years ago to study the university question. It consisted of the: Bishop of Bristol who was its chair; Kent from Physiology who served as its secretary; the Prinicipal; Young from Chemistry; Ferrier from Engineering; Barrell from Mathematics and Rowley from English.

Morris, when meeting with Kent for the first time, asked him what had happened with this committee. He answered by saying that he had forgotten but would look up the minutes of the meeting and that Morris should soon return after he had done so. In looking at these notes, Morris found that the committee had held one meeting. In this meeting, the Bishop had said that the first thing to do would be to appoint a university registrar and to create his office. This would cost £1000 per year but could not happen at the moment. Morris, in looking at these notes, felt that the majority of the committee regarded the university simply as an examining body.

Early in the term, the first tea for the members of the Joint Faculty of Arts and Sciences that Morris was at had the Principal read a letter he had received proposing a joint scheme for examination with the University College at Southampton. Though not discussed at this time, this idea was brought up again at the next formal meeting of the Joint Faculty. At this meeting Morris asked

> "Was it the opinion of the Faculty that the policy of this College should be such as to pave the way for a composite university, including Southampton and Reading, or was it their opinion that we should concentrate effort on a scheme for a University of Bristol"?

Members of the Joint Faculty had thought little, if ever, about this question but felt that a scheme for a University of Bristol was more likely. The examination scheme was not favored at all and was set aside.

It is of interest to here note that though the Joint Faculty selected two of its members to sit on the council, the staff had no share in the actual management of the college, nor in the direction of its policy. Nonetheless, one of these seats belonged to Rowley. Rowley, a popular scholar and Professor of English, had occupied this seat since the college had been founded in 1876. He was though a dyed-in-the-yarn Oxford man who held the opinion that a university for Bristol was an absurdity though he knew nothing at all of the modern British university movement. The second member elected by the joint faculty was a Mr. Francis Fry, who owned a chocolate firm. Though he was a liberal supporter of the college, he was personally a stranger to most members of it.

A further item of fascination was the lack of a senate. Though specified by the constitution of the college, this body, which should have consisted of three professors from the three faculties, had apparently never met.

Raising the university question at once brought to light that there were was a liberal and conservative "party" within the Joint Faculty of Arts and Sciences on this issue. The liberals had a definite policy that being of promoting the establishment of a university in Bristol and carefully considered every move beforehand. Morris, as their leader, had Ferrier, Francis, Ord (German), and Kent with him. Chattock (Physics) was also liberal but was relatively inactive on this issue. The Principal generally supported the liberal "party" cause, though he held the view that the views of the more senior members of the staff were to be more considered, and was almost always defeated when in opposition to it.

The conservative "party" consisted of Barrell, Rowley and the professors and lecturers primarily in the arts subjects. These individuals were men who stuck to the idea that only degrees conferred after residence at Oxford and Cambridge or in one of the Scottish Universities, or after examination by the University of London were of any value at all. They did not know that the degrees conferred by the new universities established in Manchester, Liverpool, Leeds, Birmingham and Sheffield were already considered as proof of high academic attainment. They also did not think that a degree conferred by a University of Bristol would have any value. Morris tried mightily, without success, to change the minds of these individuals.

This division, amongst liberal and conservative factions on the university question, was also true of the Medical Faculty of which Morris was a member. Fawcett (Anatomy), Kent and some of the younger physicians and surgeons who held college appointments as professors and lecturers, comprised the liberal group. The able and fair-minded Markham Skerritt chaired the conservative group. Interestingly though, the Medical School, with its long and very distinguished record, was much more ready for conversion into a university school than was the Arts Section of the University College.

It was also clear from the start that no university charter would be granted to any institution that did not include *both* the University College and the Merchant Venturers Technical College. Liberal minded individuals supported the former. The latter had been founded in the 15th century, and was largely maintained by the Merchant Venturers Society of Bristol. Membership in this organization was exclusively conservative. It included a boys' and a girls' school, an adult education division and what was best termed a university department whose well-educated students, mainly in chemistry, entered for examinations for the London University.

It was clear from the beginning that the views of the Merchant Venturer's College and the University College on the university question were totally different.

J.W. Arrowsmith

Near the end of February, Morris was invited to play a round of golf at Failand by J.W. Arrowsmith, the head of the well-known publishing house. Even though he had a proprietary interest in the University College [his firm did all the college's printing], the institution was very important to Arrowsmith. He had served on the council for many years and was a member of most of the college committees and, in addition, he was the founder of the University College Colston Society.

After finishing the round of golf, and while driving home, Morris asked Arrowsmith if he could tell him anything about the university movement, of which Ramsay had spoken. Morris had evidently touched on a sore point; Arrowsmith's reply was curt:

> "Do you think we are fools on the Council [consisting of Arrowsmith, Lewis Fry, Francis Fry, George Willis and J.P. Worsley]. We have had to exert ourselves to the utmost to keep the College on its legs, and you young chaps come here and think you can teach us how to run the place".

Morris's replied, in part, that the success of the Liverpool scheme had been due to the intimate relations of the council and staff. Arrowsmith replied:

> "The Council cannot bother about the opinions of the staff. If they want opinions, they go to the Principal. The staff have to do as they are told".

Though this statement was distressing, Morris soon learned that to get along with Arrowsmith, one had to fight back. Soon after the round of golf, Morris again raised the university question. Arrowsmith replied in the same vein again—"Do you think I'm a fool?" Morris replied back—"I'll tell you when I know you better, Mr. Arrowsmith." This brought a scowl from Arrowsmith and then a burst of laughter.

At another time, they continued their verbal parrying. Arrowsmith said to Morris—"I found a student working in your laboratory after eight o'clock last night." Travers answered—"Well, the student had my permission to be there, but you hadn't." This again brought about a laugh, which allowed Morris to talk about laboratory work and how experiments could not always be finished during the day.

Fairly quickly, Morris realized that Arrowsmith greatly cared for the University College. Morris had a special fondness for this man and viewed him as his "dearest enemy".

Fellow of the Royal Society and Helium (Re-visited)

In 1903, Ramsay put Morris's name in as a candidate for election to the Royal Society. He told Morris but Travers figured that he would be elected at some point in the *future* and thought no more of his current nomination.

On the 25th of February 1904, Morris upon getting home from the college found a telegram from his father—"Hearty congratulations, your name is in the fifteen, Travers." Morris replied back—"Thanks awfully but what fifteen." Later that same day while at dinner with two friends a further telegram arrived from his father noting that Morris had been elected to the Royal Society. Soon afterward, Morris was congratulated by his colleagues at the University College and by its council [Lewis Fry wrote him a very kind letter, the first Morris had from him].

Ramsay wrote a letter to Morris that began with:

> "Hip, hip, hooray! This is the first time in my recollection that a man has been elected the first time off. So you are highly honored. I wired your people last night, of course they are much gratified. Your mother wrote me this morning . . ."

The letter continued with

> "I have a lot of very black ones (x-ray tubes), several years old, that I got from a doctor. Soddy and I are going to heat them; and to see if we can get anything out."

This was in reference to the first experiment that Morris had carried out in Bristol which involved powdering the blue tinted part of an aged x-ray tube, heating the glass in a combustion (hard glass) tube, collecting the gas, and after treatment removing all but the rare gases and then examining the residue spectroscopically in a very small Plücker tube. The helium spectrum was seen, faintly but distinctly. After doing this, Morris had written to Ramsay.

Morris did not recall what theoretical reasons suggested this experiment or what conclusions he drew from it. He though wrote up a short paper and showed it to Ramsay who said that this work was nonsense and that if Morris published the paper he, Ramsay, would say so in print. Morris consequently put this paper away and forgot all about it. Later though, in 1911 or 1912, Ramsay, having forgotten all about the previous incident, carried out an identical experiment, obtained a trace of helium, and published a paper with the Chemical Society, suggesting that the helium was produced in situ by the action of the electric discharge. At the meeting where this paper was read in 1913, Collie and Patterson read a paper in which they described experiments showing that helium was obtained by passing an electric discharge through hydrogen, and that if oxygen, free or in combination, were present, neon was also formed. These papers created a furor and the authors received immense, though temporary credit for their discoveries. It was, of course, shown later that the rare gases found were derived from the air.

University movement and other affairs (1904-1905)

Late in March 1904, Morris met with Ferrier, Francis and Rafter over dinner to discuss the best way to push the university movement forward. They decided to ask Kent to summon the committee appointed by the council on the university movement and that Morris should be asked to attend the meeting in place of Young [who was of course no longer at the college].

In April, Morris went for a week to Wootton Fitzpaine, near Charmouth, to stay with Alfred Capper Pass, a Bristol industrialist and friend of the Ramsays, who had formerly been a member of the college council. His business was in the recovery of metals such as tin, lead and copper from scrap and waste. Pass was a man of great ability and wide interests and an ardent field geologist and archeologist though he was an invalid when Morris visited him. He was to die in 1905. His son Douglas later endowed a chair in organic chemistry at the university that still exists to this day and is named for his father.

After this visit, Morris went to Rhyl for his brother Cecil's wedding. He also stopped in Liverpool for a day to see friends he had recently made and to put in a further good word for his friend Donnan and his attempt to obtain the chemistry chair position at the university.

May saw Morris and Ferrier spend some time writing a report on the proposed university. This draft report was an outline of the advantages which higher education would gain from the establishment of the university. It was followed by a draft constitution, which was similar to that for the other recently established universities in England.

Ramsay, with whom Morris had shared this draft document, gave constructive advice on it and noted that he was joining a deputation to help ask for larger grants to universities. He further told Morris that he believed that there was a good chance for his group's success in this endeavor but it would probably not affect Morris, since he was not yet at a university. This additional money for the universities might though be, according to Ramsay, a lever for Morris and his colleagues to try for one. He also congratulated Morris for having moved the university cause so far for Bristol.

The draft report was submitted to the committee of which the Bishop of Bristol was the chairman and also included Kent, the Principal, Ferrier, Barrell and Rowley. This body, now reconstituted as a committee of the council and senate, sent the report along to the senate, which had been reconstituted at Morris's suggestion. The senate consisted of professors of all three faculties, and was representative of the whole college and much less unwieldy than a meeting of the whole of the three faculties.

In June Morris attended, as a guest, Old Boys' Day at his old school, Blundell's. At this happy event, he had the opportunity to discuss the Bristol university scheme with a very enthusiastic old Blundellian, Lionel Goodenough Taylor, Editor of the Bristol Times and Mirror. Taylor's son and Morris had gone to school together and were friends. This contact was to be a useful one.

In October 1904, Morris had the pleasure to visit Donnan in Liverpool. Donnan, who had been selected as the new chair of physical chemistry, was in the process of building his laboratory there. Morris lectured that night to a meeting of the Student's Chemical Society and later had a late and merry evening at the University Club.

In January 1905 Morris had the opportunity to go Glasgow to consult with the Brin Oxygen Company [now a part of the Linde Group] on how to properly use their newly installed air and hydrogen liquefiers that he had helped design for them.

He arrived on the 11th and had a sufficient quantity of liquid air available to attempt to run the hydrogen plant on the 14th. Five separate attempts on this day to liquefy hydrogen were not successful; each time this was tried it failed due to impurities present in the system. On the 6th and final try of the day, with Lord Blythswood (originally known as Archibald Douglas Campbell—a "gentleman-physicist" who, in his home laboratory at Renfrew, had almost discovered X-rays several years before Roentgen) present, they made a half-liter of liquid hydrogen—much to the delight of the Lord.

CHAPTER VII

Bristol University—Problems

Meeting with Haldane

Near the end of 1904, Morris felt that no real progress was being made with the university scheme. Looking for suggestions, he spoke with Ramsay and Ramsay offered to introduce Morris to the Liberal politician R.B. Haldane [who would later become an important part of the British government—including, in part, being Secretary of State for War from 1905-1912]. Morris then wrote asking for an appointment at Haldane's London office. This meeting was set up for the afternoon of January 17th, 1905.

Haldane had been sent an abstract of Morris's report to the college senate and council. He sat Morris down in a chair by the fire, gave him a cigar and began to talk. Ramsay had told him all about Morris and his efforts to obtain a university for Bristol; Haldane made no enquiry as to Morris's official status, likely because he had none himself since his party was not in power. Given this, Haldane noted that he could not be directly involved in any negotiations between the college and the government but his advice would certainly be asked before any final decision was made. He though stated that he could speak authoritatively as to the conditions that must be fulfilled before a university charter could possibly be granted for Bristol.

Haldane noted that the day of composite universities had passed and also agreed with Morris's ideas as to internal examinations that he had put forth to the college committee. Haldane continued by insisting that the court could perform a real function by keeping the university in touch with the public and that Morris and his colleagues should consult other universities and their constitutions.

Two main points that had to be considered were how to approach the Merchant Venturers Society and how to raise money for the university. It was obvious that the university must include both the University College and the Merchant Venturers Technical College. Unfortunately, the differences between these bodies had only intensified since their supporters were of opposite political parties. This had to be overcome and had to involve resolving differences in general principles between the two factions.

To begin accomplishing this, Haldane suggested Morris should see Walter Long [later to be Secretary of State for the Colonies and First Lord of the Admiralty] who currently was a Conservative M.P. for North Wilts, Sir Michael Hicks-Beach [former two time Chancellor of the Exchequer and later known as Lord St. Aldwyn] and the Duke of Beaufort. Morris should have these men form a committee external to both institutions that should primarily be made up of Conservatives. This committee should approach both institutions and ask them to accept the principle of co-operation in the formation of a university. The institutions could settle details subsequently. The committee might then form the nucleus of a body including the Mayor and citizens of Bristol that would begin raising money for the university.

Aftermath of Haldane Meeting

Morris had not told Lloyd Morgan that he was meeting with Haldane. He was certain that, if he had, he would not have been allowed to go or discouraged from going. He also knew that though Ramsay would have met with Haldane while he was Principal, Lloyd Morgan would not do so. Morris also feared that the meeting with Haldane would cause higher authorities in the college to react very, very badly toward his continued employment in the University College at Bristol.

Though he realized he was taking a very significant risk, Morris felt it was worth it. He also thought there was no logical reason why he should not seek unofficial advice on a subject he was interested in from the leading authority on that subject. His meeting also did by no means commit the council or the college to any action whatsoever.

Having done the deed though, Morris decided to do the right and honorable thing and tell Lloyd Morgan about his meeting with Haldane. Morris began this discussion by telling Lloyd Morgan that he was not at all intruding on the relationship of Lloyd Morgan and the council of the college and that he had no intention of going further until Lloyd Morgan could speak with Lewis Fry whose family was widely known for their successful chocolate business.

Travers thought it odd that Lloyd Morgan was more upset that Morris might get into serious trouble with the council than by Morris's intrusion into an area that was Lloyd Morgan's alone. Lloyd Morgan did see Fry a day or two later but told him nothing about what Morris had done.

At the end of January 1905, Morris called on Fry and told him the whole story. Fry listened very gravely and said nothing until Morris had finished. At that point, he paused and said

> "Well, Professor Travers, you have put me in a very difficult position in which there are only two things that we can do; we can dismiss you, or we can go ahead, and I think we had better go ahead."

Fry never turned back and became the active leader of the university movement. Morris felt that it was his good fortune to be associated with him during the next year and a half and that this association was one of the happiest in Morris's life.

The committee of the senate and council, originally the Bishop's committee, met on February 8th. This meeting, likely called by Fry himself, had Morris report on the events of the previous month and the whole university matter was discussed. A decision was reached at this meeting to begin outside negotiations and get in touch with local magnates. A group consisting of the Bishop, Fry, Dr. Ernest Cook (the Chairman of the City Education Committee) and Morris was formed to do this. Morris immediately asked Mrs. Napier Abbot to write to the Duchess of Beaufort who, it was thought, might give more assistance than her husband the Duke. He also made contact with some members of the City Council of Bristol on the university matter.

Soon afterwards, Morris wrote to Haldane. A portion of this letter follows:

> "The Duchess . . . . . has been approached through a lady who is very keen about the matter, and in her reply, and in a letter to Mr. Lewis Fry, she has expressed herself most cordially. She is anxious to take the matter up, fully recognising the importance of excluding politics, and believing in the possibility of carrying it through.

> I have used your name in the matter so she may write you. In any case a line from you will confirm the matter.

> I have also laid your views and suggestions before a small private meeting of members of our Council and Staff. Among them was Mr. Lewis Fry, Chairman of our Council, and Dr. Cook, Chairman of the City Education Committee. Both of these gentlemen are keenly interested in the movement, and are in perfect agreement with you as to the course of action that should be taken . . .

> Those of us who are concerned in the matter realise that we must act carefully, so as to avoid coming to a deadlock with the M.V'S, and this might result from the impression that we are merely working for our own college."

Morris had also, in this letter, asked Haldane to write or speak with Sir Michael Hicks-Beach and Walter Long about the university scheme saying that Fry would be prepared to go to London to meet them. Haldane wrote Morris back stating that he had already spoken to them; but Fry was ill, and he had asked that Morris stand in for him. Morris thus made two more visits to London, first meeting J.W. Bull, Long's Parliamentary Secretary, and then later Long himself. Long said that he had played a minor part in the foundation of the Liverpool University and would be very glad to help. Long further noted that he would see Sir Michael Hicks-Beach and use his own influence with the Merchant Venturers Society to help.

By the end of the spring 1905 term, Fry had also obtained promises of support from Lord Bath, Lord Ducie, Sir John Dorrington and Lord Lansdowne. The Bishop of Bristol had also already promised his support.

Of interest was the fact neither Lloyd Morgan nor Arrowsmith took part in any of these discussions. Neither of these men had much, if any, faith in the university movement and Lloyd Morgan felt that since the committee that was to be formed would be external to both colleges; he would be best to stand aside on this matter. Arrowsmith was fully engaged in the necessary and excellent work he did in raising money so that the college could continue functioning. Recently, Arrowsmith had achieved the triumph of the completion of the buildings on College Road; that would become University Road in the future. The press, which knew nothing at this time of the university movement, termed Arrowsmith's accomplishment as the completion of the University College.

Personal matters and Students

Morris was at this time paying, from his own pocket, the costs for his travel and postage on behalf of the university scheme. He also had significant expenses attached to his appointment as chair of chemistry. Since the Chemical Library only subscribed to the Journal of the Chemical Society and the German Berichte der Deutschen Chemischen Gesellschaf, Morris felt that he had to purchase, for the college and it students, the Chemisches Zentralbatt and the Zeitschrift fur Physikalische Chemie. The college also made no contribution at all to the costs of research incurred by Morris and his students. However, Morris though considered himself fortunate in that the college had received an annual gift of £1000 from Fenwick Richards; this allowed Morris's own salary to be paid in full.

Fortunately, Morris was unmarried and living was relatively cheap and simple. He did not have a car and was quite content to have a bicycle. His vacations were simple and frugal; he could spend five weeks on the continent or in Norway fishing for as little as £25. His entertainment was inexpensive and involved little more than whiskey or port and good cigars and/or friendly little dinner parties with games of whist afterwards.

As far as friends were concerned, Ramsay's friends were good to Morris but they were a different, older generation than Morris and his association with the Waits, Tyrons, Frys, Passes and others, though quite pleasant, was quite formal.

Morris was fortunate though that Francis introduced him to his circle of friends. These included some very pleasant and interesting people, including: John Curtis and his wife and Robert Carpenter. The Curtis's Saturday evening whist parties were delightful, even though Morris felt he played a poor hand, while Ferrier and Francis were better. At these events, 11 P.M. always saw a large plum cake appear with drinks. Sundays typically had a group take a long walk, sometimes to Cheddar and lunch on bread, cheese and beer usually at a pub by the Brendon reservoir. After a return to Bristol by train, supper with Carpenter, who lived with his sister, was common.

Morris also often had supper with his senior students on Sundays. These students included, in part: F.L. Usher, who went to India and afterwards became a lecturer in colloid chemistry at Leeds; F.P. Burt, who became a lecturer in Manchester; A.G.C. Gwyer, who became chief chemist to the British Aluminum Company Ltd. and Miss Williams who had been working on the chemistry of cooked fish since Ramsay's time at University College.

Morris could overlook an occasional absence by a student to one of his lectures but students who did not attend regularly were, when present, generally inattentive and a nuisance. Making oneself a nuisance was a major crime to Morris and was very, very difficult for him to deal with. One occasion saw two excellent young men miss a whole week. These individuals had argued over the Dare girls, who performed in the Bristol pantomime, and the argument had led to a fight. They both ended up so battered and bruised that they did not wish to go out in public for a week's time. When they returned, Morris asked them—"Mr. X and Mr. Y, you have missed four lectures. Have you been ill or have you met with accidents?" At that point, Morris and the two young men all burst out laughing and the discussion ended with Morris remarking—"You silly asses!" X married Y's sister and Y married X's sister. Interestingly enough, about twenty years later, Morris, having forgotten all about this, spent a night at X's house and after dinner told this story. X laughed deeply and said—"You have forgotten something, X and Y, were me and my brother-in-law."

A second story illustrated the mentality of the student at that time. This occurred on or about May 28th, 1905; while the great naval battle of Tsushima between Russia and Japan was being fought in Asia. Leaving University College one day, Morris saw a student reading the evening paper and asked—"What's the news?" The student responded—"Splendid, X-shire are all out for Y runs!"

City of Bristol

A question that often puzzled Morris was why University College Bristol was so poorly supported compared to the colleges in Manchester, Liverpool, Leeds, Sheffield, Birmingham, Southampton and Reading. It was of course obvious that a great deal of private money had been spent on technical training by the Merchant Venturers' Society. If it had not been for this private money, the city of Bristol would have needed to come up with this money. Even with this in hand, the position of University College Bristol was poor.

Bristol had fallen in a century from being the third largest city in England and the next most significant port after London. She had to a large extent lost the industries and trade that had made her wealthy a century before, such as the manufacturing of cloth and the export of wool. The floating harbor in Bristol City and the Portishead Dock could not accommodate modern ocean-going ships and her carrying trade had passed mainly to Liverpool and partly to Southampton. Chemical industry commenced on a large scale at Netham by the United Alkali Company Limited could not though compete with the more modern Brunner-Mond organization. The last glass house, a survivor of a flourishing industry scattered across the area, had closed down at the start of the 20th century.

However, Bristol was far from a poor city and certainly was not in decline. Its citizens had realized, in time, that they needed to renew their traditional enterprise. Evidence of this was seen in the building and opening, in 1904, of the great Avonmouth Dock that was able to accommodate the largest vessels then sailing. Manufacture of tobacco was beginning to be of vast importance. Other industries such as the manufacture of chocolate, paper products, china, furniture and fabricated steel work as well as others were growing and new factories were springing up around the city.

Unfortunately, Morris did not include this renewal of the city in his argument for a university in Bristol but based his position on examples set by the other great cities in Britain, Germany, Russia, Austria-Hungary and America and the needs of the moment. He though, later in his life, realized that 1904 was the beginning of a tide that portended much for the future and that timing was just right for a university in Bristol; further delay would have been disastrous, if not fatal, for such an institution.

Lloyd Morgan and University College

The joint faculty had the opportunity to nominate two members for the college council and for many years their representatives had been Professor Rowley of English and Francis Fry. Fry, though he had leanings toward science, was a businessman in the chocolate industry and, as such, he could hardly represent academic views and interests. Professor Rowley, as stated in this previous chapter of this book, believed there to be only two universities in England, some fair imitations of universities in Scotland and a number of so-called universities which had recently come about but were mislabeled. He did not visualize a university for Bristol.

When Rowley had been last re-elected to the council, it was understood that he would accept election but resign the honor immediately. However, it appeared to Morris and his colleagues that they might have to wait until the next council election in order to elect a representative who would advocate for the progressive members of the staff.

Surprisingly though, at a poorly attended meeting of the joint faculty at the beginning of the summer term, Lloyd Morgan opened the proceedings by noting that Rowley had resigned from the council. After the usual tributes had been made for Rowley and his service, Lloyd Morgan stated—"We will now proceed to elect a new representative on the Council in place of Professor Rowley, and I think that all will agree that our friend and colleague, Professor Barrell . . . ." Lloyd Morgan got no further when Morris interrupted with the remark—"Mr. Chairman, this matter requires notice of motion." Lloyd Morgan replied—"Yes, perhaps I had better put it down on the agenda for the next meeting."

Barrell was a good friend of Morris though he was a Cambridge man and his ideas about universities were similar to those of Rowley. He simply could not contemplate a university for Bristol and that its degrees would have any value. The only qualification that he had for service on the college council was the fact that he was the next most senior member of the faculty to Rowley.

Additional discussion of representation on the council led to Morris's colleagues proposing that he run, but Morris categorically refused to do so. Morris wanted Ferrier to be the individual selected because he was a more senior member of the staff, an older man than Morris (aged 47 vs. Morris's 33) and a bit more distanced from the issue than Morris. Morris felt that having a person who lived in his home and had similar views to him on the university movement would be advantageous. Support for Ferrier was easy to gather and the following day Morris wrote Principal Lloyd Morgan the following letter:

"Dear Mr. Principal,

In electing a representative to succeed Professor Rowley on the Council, it is our duty to consider the particular needs of the College at this critical period of its history.

If the university movement is to be carried out to a successful issue, it is necessary that the policy which has been declared both by the Council and Faculty should be kept in the forefront, and above all that such course of action as is taken shall be regulated by a knowledge of the problems involved.

We need a representative who is thoroughly conversant with the university question, with the needs and requirements of a modern university, and with its relationship to the community, and who has had the opportunity of studying the problems at home and abroad.

I have discussed the matter with some of my colleagues, and I am voicing their opinion in suggesting that in Professor R.M. Ferrier we shall find a most able representative; for beyond possessing the qualifications I have mentioned, his wide knowledge of the commercial life of Bristol, and his acquaintances with many of her leading citizens, give him a certain facility for advancing our interests."

After sending this letter, Morris had a meeting with Lloyd Morgan. Lloyd Morgan was still determined to nominate Barrell but he had an idea for a compromise solution. There was a vacancy of a Governor's nomination to the Council and he proposed that he should write to Francis Fry and ask him to resign from his faculty elected position and take the Governor's nominated position. A day or two later, Lloyd Morgan told Morris that he had seen Fry and that he agreed with this arrangement. Lloyd Morgan was to send a letter to Fry stating this and also arrange for his formal nomination and election by the needed number of Governors.

But, the day of nomination/election came about with no answer from Fry. Ferrier had been nominated; and when Morris went to see Lloyd Morgan, he told Morris that was going to support Barrell. Morris then told a horrified Lloyd Morgan that the matter would be fought out during the imminent joint faculty meeting but stated good-naturedly that he would not continue this fight outside of this body. However, a telegram from Fry, arriving minutes before this meeting was to begin, formally agreed to the Governor nomination. This compromise satisfied all.

Morris's Research on the University

For a long period of time, Morris had spent significant time in reading the reports of the American Commissioner for Education and Britain's Education Department. Morris learned a great deal in doing this. This information dealt with developments in a variety of countries, including France, Germany and in England.

French universities began with an act of Napoleon who established in 1803 the seventeen provincial academies, which were subordinate to the University of Paris and subject to a central examination system. The object for this was of course political and was done away with in 1896 where seventeen independent universities were established. The university system was progressing well in 1905 but was to be greatly hindered due to the suffering France experienced from World War I and the curtailment of university grants.

In Germany the university system grew rapidly after the battle of Jena-Auerstädt in 1806 as well as their influence as centers of national effort. The Peace of 1815 led to ideas of national reconstruction and a review of education on all levels.

Simultaneous movements resulted in the foundation of Mechanics Institutions in 1823, University College in London in 1826 and of King's College in London in 1828. Interestingly enough, University College London was due to the poet Campbell who wanted to establish, in or near London, an institution that was secular in nature. This was met with fierce opposition and the result was the almost simultaneous foundation of King's College by supporters of the church.

Proprietors of a University College and the press called their institution the University of London. It seems to have been anticipated that there would be no difficulty in getting a charter for the college under this title, so long as it did not give it the right to confer degrees. In 1837 an independent corporation with the title University of London was established with University College and King's College as its constituent colleges. Candidates for the qualifications, which it conferred, were required to study at one of these two institutions.

This system lasted until 1858, when partly as the result of claims by other colleges to be admitted as constituents and partly with an idea that the University of London should be Imperial, the 1837 charter was revoked and the university became a purely examining body.

Many other factors influenced the government in taking this action and it may be that the basic idea was examination by a public body eliminated patronage and secured the entry to the professions and to the civil and military services of young talented men. Certainly the results of the Civil Service Examinations, and those for admission to the Woolwich and Sandhurst military academies justified themselves; but the system restricted the candidature of young men who had followed a classical course of education in small or large public schools for civil service appointments.

The University Colleges had developed either through the efforts of private individuals as was so with Owens College in Manchester or were established by groups of citizens with the object of extending cultural education locally as was so with the University College in Bristol. The existence of the London University, which could confer degrees on their students by examination, encouraged these institutions to raise their standards.

But it was not long before these institutions felt restrained by the University of London and took action. In April 1880, Victoria University was chartered at Manchester; University College Liverpool joined this institution in 1884 and Yorkshire College in Leeds became part of Victoria University in 1887. In 1893 the Welsh University Colleges also became independent of London. 1903 saw Liverpool leave Victoria University to become the University of Liverpool and 1904 had Leeds receive its Royal Charter and become the University of Leeds. Other University Colleges that became chartered universities were Birmingham in 1900 and Sheffield in 1905.

A long struggle in London ultimately led to the division of its university into two sections. An external section still conferred degrees on all comers by examination only. But the London Colleges and Medical Schools became schools of the university. Candidates for internal degrees were obliged to follow a course of study in one of these schools and passed examinations, which were basically controlled by their instructors.

The system for giving general examinations had been based on the desire to eliminate patronage and a fear that control by instructors might lead to standards being lowered. It was evident though that academic freedom did not involve lower standards in German universities and in the modern American universities. It had also become common that young men interested in scientific careers took a bachelor's degree in England and then spent time in Germany to earn a Ph.D.

degree after presenting a thesis on some research topic. How odd it was that an Englishman should have to go to Germany to find freedom from central control in what he wished to study!

No one understood this situation better than Ramsay and no one, virtually alone, fought harder to correct this problem. He felt it the duty of instructors in London to have the dominant role in putting together their courses and in examining their students. At the first meeting of the Board of Studies in the newly constituted London University, the topic of "internal" examiners came up for the first time. Ramsay, who with Morris was at this meeting, dealt with the revulsion that most present felt about this idea by stating—

> "If the teachers refuse to accept the responsibility which is now theirs, and to take a part in the examination of their own students, much that we have fought for years to win will be lost."

At a meeting of the academic members of the committee of the council and the now re-vitalized senate on May 9th, Lloyd Morgan had suggested that a "prospectus" should be drawn up that would describe the history and aims of the college and of the prospective university. Lloyd Morgan offered to do this but Morris, who had already done the research necessary for such a task, ended up writing it and termed it "the pamphlet".

Work on "The Pamphlet"

Morris spent a great deal of his remaining May and early June working on the pamphlet and showed it to Lloyd Morgan in the middle of June. Lloyd Morgan suggested that it be typed up and sent with a cover letter to Lewis Fry which stated, in part:

> "In drawing up this note I have kept Mr. Haldane's remarks of last January very clearly in mind,—'Get the two bodies (U.C. and M.V.T.C.) to accept the principle of co-operation in the construction of a university before you begin to consider the details,'—I am convinced that he is right, and that if we begin by considering what work each of these schools shall take we shall get into difficulties.
>
> I think, too, that the City Council should be approached at the same time as the U.C. and the M.V.'s, and that that body should be mainly instrumental in deciding what sort of a University Bristol is going to have. I feel that the U.C. can trust itself in the hands of such a committee as I suggest.
>
> I am glad to hear how well things are going."

With regard to the words, "what sort of a University Bristol is going to have", Morris noted that too many people referred to an institution like that in Manchester, Sheffield or Birmingham. The objective though of the University College in Bristol had originally been educational and cultural; and that the professional side had developed to meet local needs. The aim of Morris and his colleagues was for a university that would not only serve the city of Bristol (population about 360,000) but also serve the west of England (having a population of perhaps 3,000,000). They had to meet not only the cultural needs of a large area but also the agricultural, industrial, commercial, professional and intellectual needs of that same area.

On June 19th, Morris wrote to Haldane to report on progress to date. He noted Lewis Fry was "most anxious to take the matter up." Morris knew that he should take a step back and let Fry approach the various important people whose support was wanted [and needed] for the university.

The need for a site for the university was also of importance. The Merchant Venturers College could not be added on to so a new location was critical. Fry was aware of this and was working on acquiring the two acres of the Blind Asylum ground, hopefully for £25,000. This location actually bordered the college. Ferrier, thinking ahead, had arranged with the city Surveyor to make a plan of the block including the College, the Blind Asylum, the Museum and the Art Gallery. Six months later, the Blind Asylum site had been acquired.

Morris concluded his letter to Haldane:

> "It is difficult to impress the fact that a big scheme should be put forward, as one is always told that it will be quite impossible to find the funds. Unless Bristol differs from every other great city in the country this will not, I believe, be the case.
>
> ' . . . . if you have any suggestions to make I shall be very grateful for them."

He saw Haldane a couple of days later at a Royal Society event where Haldane told Morris that he was pleased with him and what was occurring with the university movement.

India—Revisited

Forces were already at work, and had been for some time, that would ultimately take Morris away from Bristol in 1906. This all began back on July 9th, 1900 when Morris was weighing a bulb of pure neon, which allowed the density and atomic weight of this element to be determined. Ramsay, sitting beside Morris, began discussing his upcoming visit to India. This was at the invitation of the wealthy J.N. Tata who proposed to endow a research university and wanted Ramsay's advice on how to organize it.

Ramsay asked Morris if being the first Director of this entity would appeal to him. Morris immediately refused and Ramsay did not press the matter further. In retrospect, this was the first Morris had heard of this; he had not talked much with Ramsay the last few weeks. In addition, Morris was ill enough (with the aforementioned gastric ulcer) that the work at hand was extremely difficult on its own and he actually wondered if he would ever be healthy again.

In October 1901, Morris was asked to meet Tata at a dinner that Ramsay was hosting but Morris was again too ill to attend. Ramsay, in March 1904, wrote Morris about the proposed Indian Institute of Science (IISc) and sent him a pamphlet about it. Still, Morris had no interest in this undertaking. September of 1905 saw Ramsay write Morris again about the IISc and send him another pamphlet about it. In his reply to Ramsay, Morris wrote—

> "I have at present staked my money on Bristol. I should like to see the University on its legs, and then if the place still remains among those of the second class, I shall have a chance for one of the more important posts."

"The Pamphlet"

Before he left for his 1905 summer vacation, Morris arranged to have 50 copies of his pamphlet typed up. In July of 1905, Morris also resigned his position as honorary secretary of the committee, formed to deal with the university matter, but he still remained a member of the committee until May 1909 when the charter for the university was finally granted.

Early in October, he sent copies of the pamphlet to members of the committee of the senate and council and also to the City's Education Committee. The City Council asked Morris if they could circulate the pamphlet to their members; this he readily agreed to.

At the University College, several meetings were held of the joint committee and were adjourned to give some members, particularly professors of the arts, more time to think over Morris's proposals. They formed the majority of the senate and on October 24th, it was clear that the senate, as a body, would oppose the idea of establishing a university at Bristol. Their main criticism of the university scheme dealt with Morris's refusal to put the educational systems established in Oxford and Cambridge above all others. Though obviously disappointed at this turn of events, Morris was somewhat heartened by a remark from Lewis Fry that he was by no means discouraged.

The next day, Morris was surprised to receive a telephone call from Arrowsmith who said:

> "I want to congratulate you on your pamphlet, it is first rate. I have shown it to several members of the Council, and we have decided that it must be printed. Come and see me."

The people he named were also friends of Ferrier and Morris and were all well acquainted with the university movement. Arrowsmith, who had been at the meeting, still was prejudiced against the university movement and had obviously not carefully read the pamphlet. After this meeting on the 24th he had gone to the Liberal Club for tea and met the men he noted who helped him form his opinion of the 25th.

Given this, Morris called on Arrowsmith at once. He told Morris that he proposed that the University College Colston Society should print the pamphlet and distribute it at its dinner in March 1906, where he would be Chairman.

Following this, Lewis Fry and Dr. Ernest Cook met the new Master of the Merchant Venturers Society and impressed upon him that those involved with the university movement were not doing this only for themselves but for a University for Bristol, which must include both institutions.

Morris also began a correspondence with Professor Alfred Marshall who was the first Principal of the Bristol University College from 1877-1881. Morris found that Marshall had no knowledge of the modern university movement and still held the idea that the London degrees had a much higher value than any obtained outside of Oxford and Cambridge.

His second letter to Marshall noted,—" . . . You write me objecting to the modern university on principle, and hold up the University of Cambridge as a model of unattainable perfection." Morris responded by noting that neither Oxford nor Cambridge had in 1905 a school of chemistry worthy of the name. Marshall writing back referred to the "brilliant schools of physics and physiology in Cambridge." Speaking generally of university education he also wrote—"I fear I put the ideal higher than you." Morris replied:

> "Is not this somewhat gratuitously unkind. If you will read my pamphlet, and Ramsay's (introductory) letter carefully . . . you will, I think, arrive at a fairer conclusion."

Marshall had also spoken of the risk of hasty action to which Morris answered:

> "Further, I am only in a hurry to make, not only this College, but the educational forces of Bristol, pull in the direction which will lead to our ideal, though that ideal may not be attained till after many years."

Marshall had also sent to Morris a newspaper piece to which Morris remarked—"A defence of Cambridge by an anonymous writer appears to me to be an insult to that university." As a result, it was obvious Marshall did not want to modify the scheme for a university; he wanted to utterly destroy it.

CHAPTER VIII

Bristol University—Problems Resolved

Morris as an Administrator?

The end of October saw Morris meet with Arrowsmith at the golf club on a Saturday afternoon where Arrowsmith said how unhappy he was with the progress of the university movement and that he wanted Morris as Principal. Morris, telling him he would think about this offer, noted a few days later he would be willing to be Vice-Principal for one year, with an extension of office if need be, as long as he would be compensated for losses in his other work. Morris was not interested in the title but realized that such a position would help the university movement more than a mere professor's position would. Lewis Fry also thought this to be a good idea.

Morris wrote to Ramsay and let him know what was occurring and he replied warning Morris that he might take the wrong path [the Vice-Principal position] and asked if he had spoken with Lloyd Morgan about this. Travers could not do that since he did not really want the position and it was not his idea to begin with. Morris wrote back to Ramsay on November 15th saying:

> "I don't think you quite understand me. I would gladly be rid of the whole thing; and I am only giving my time to this work because there is no one else to do it.
>
> My interviews with Fry and Arrowsmith were the result of repeated suggestions that the Council, or rather the active members of it, looked to me to take charge of this matter. A. had said to me the day before 'I wish to goodness we could make you head of the College, '—I would not take the post in any eventuality, And I thought we must look forward to the time when we could appoint a Principal who would not be a member of the teaching staff.
>
> In my letter to Mr. Fry I made it quite clear that I would not give up my original work in science, and that I should take up the university work at a

sacrifice, and for a limited time. However, I would rather do one thing at a time, and that efficiently.

Then with regard to the money question, my salary is £350 a year which I can increase by outside work, writing, etc. I have no private income. If I go on with the university work, I increase my expenses, and lose all chances of making more than £350. I am not exactly grasping; but I am worth more than that.

My position is this. The university scheme of which L.M. and Young told me, was nothing more than a dream. Whatever exists at the moment, *I have made . . .*"

Morris also noted that to give up the university movement and put it back in the hands of Lloyd Morgan would be disastrous, if not fatal for it.

On the following day, there was a general meeting of the college. At this meeting, the Bishop of Bristol spoke of Morris's pamphlet as setting forth the aims of a modern university and called on the rich men of Bristol to support the scheme. Several members of the council said that the pamphlet had converted them to now favor the university movement; Morris then wrote formally on November 21st to the Secretary of the University College Colston Society and made his pamphlet available to them.

The Blind Asylum

In December several dramatic events occurred. Previously, Morris had written to Haldane on June 19th about the possibility of acquiring the Blind Asylum site, adjoining the college buildings, and with a frontage on the top of Park Street, for £25,000. New quarters were not only needed for the university to expand into but also for it to have a more prominent stature and location. Ferrier had already arranged with the City Surveyor to make a color plan of the block including the: college; museum; art gallery; drill hall and the blind asylum. The council was not enthusiastic about this though. In fact, Worsley said, when meeting with Morris and Ferrier,—"You might possibly get the playground, but the frontage is out of the question."

However, serendipity intervened. Morris, making a phone call one day found the line busy. Since the college shared a telephone connection with the Blind Asylum, Morris, before hanging up, overheard a few words of another conversation. Normally, he would have hung up and tried his call later but he listened further and learned that a sale of the Blind Asylum property was being negotiated! Morris immediately went to Fry and told him this; Fry was astonished at not having heard anything of the matter. He would ask around regarding this.

A day or two later Fry sent for Morris. Fry said that what Morris had told him was quite correct; the site was for sale and a possible buyer was negotiating a purchase. Fry said—"But what can we do?" Morris responded with—"You, Mr. Fry, can do one thing. Ask your cousin Mr. Joseph Storrs Fry to buy us the option. If he were to do this, I don't doubt that we can raise the balance." Fry paused and then said he might.

Morris was quite worried about personal matters [his father William was very sick] and he went to London and stayed with his parents for three days returning to Bristol the day after Christmas. In fact, at the moment of leaving home, William shook hands with Morris saying—"Godbye, God Bless you, my boy." Morris felt that this was their last goodbye. Though he did not know then, it was his last Christmas in his father's house.

Upon returning to his home in Bristol, Morris found a note from Fry asking for Morris to call him. On the 27th, Morris called Fry who asked that Morris see Lloyd Morgan and that they draw up a report on the Blind Asylum buildings after they had carried out an inspection on them. After looking over the buildings, Lloyd Morgan left writing the report to Morris who sent two copies of the report and a plan of buildings to Fry on the 29th. This report said that while they might be able to use the buildings, there needed to be some minor renovation work, though this was not completely true; some of the facilities were in desperate shape—but those involved in the university movement wanted the ground and frontage property so badly that they embellished the truth.

On January 11th, 1906, Fry told Morris that he had raised the needed money [and more]. This money was from: Storrs Fry £10,000; Sir W.H. Wills—£10,000; Francis Fry—£5,000; Sir F. Wills—£5,000 as well as some smaller sums from others. Fry told Morris that he and Lloyd Morgan were to see the Lord Mayor and that the Bishop was to write to Sir George White and Lady Smythe. Given these very generous donations, Morris felt it necessary to draw up a definitive scheme; too many people were mistaking his pamphlet for a scheme when it was really a sermon!

University College and Vice Chancellor Position

On the 14th Morris played golf again with Arrowsmith and they discussed the position of Vice-Chancellor at the new university. Arrowsmith said that the Vice-Chancellor should be an administrator only and that he wanted the College Colston Society to finance the university movement and appoint Morris as secretary with an honorarium. Independent of this meeting, Morris was aware that Lloyd Morgan himself wanted to return to being a professor and that if Morris was to be formally considered as Vice-Chancellor, Lloyd Morgan would oppose him and also enlist the assistance of other "old-timers" from the university college staff to do so.

Morris's friends in Bristol also advised him on the university scheme. These individuals included: Ferrier, whose golfing ability (he had a handicap of only two) gave him entrance into a wide circle in the golf club and in the Clifton Club; Robert Carpenter who was a senior partner in a leading law firm; Hall, an importer of fancy goods whose mother had been a significant international journalist; Nixon, a young doctor who would become a professor of medicine in the university and George Crawford who was a military man. These people were among Morris's closest associates and his sounding boards.

Merchant Venturers Society

With the exception of David Robertson, who would later become a Professor of Electrical Engineering in the new university, Morris knew none of the staff at the Merchant Venturers Technical College (M.V.T.C.). Morris had previously called on the Principal of the M.V.T.C., a man named Wertheimer, in January 1904 and invited him to dinner with he and Ferrier. However, that invitation was declined in a very cold, formal manner.

In December 1905, Lewis Fry, Lloyd Morgan and Ernest Cook called on Russell Harvey, the Master of the Merchant Venturers and asked him if he would be on the committee that was being formed for the purpose of approaching the governing bodies of the two colleges. Harvey noted at this meeting that he was favorably inclined but must consult with Pope, who was the member of their body who dealt with the affairs of the M.V.T.C. Pope though refused to move until he knew what exact scheme was proposed which came as no surprise to Morris. In the summer of 1906 there was a second meeting [with Lewis Fry, Ernest Cook and Harvey present] that clarified some misconceptions about the university scheme that Harvey had and led to the establishment of better relations between the M.V.T.C. and the university college.

Wertheimer, who was a shrewd man, realized that the university was coming and that he was in a difficult position and was, as a result, extraordinarily cautious. He was pleasant enough and noted that he wanted to come to terms but was not willing to discuss matters of "mutual interest".

Press Coverage

On January 23rd, 1906 Lloyd Morgan came to Morris and wanted facts on which to base news articles for the university movement. Two days later Lloyd Morgan showed Morris notes on an article in which the matter was discussed between *Advocatus Diaboli* and a university professor. It seemed that Lloyd Morgan wanted to put the article into the hands of a young man who would write what he wanted. Morris decided that he would do it instead. Interestingly enough, Lloyd Morgan suggested sending this article to the Tory paper Bristol Times and Mirror which viewed the university college as a liberal institution. Its editor happened to be Lionel Goodenough Taylor!

Taylor's son was an old school friend of Morris so Morris asked him to talk to his father, whom Morris had already met at the Blundell's Tercentenary, and tell him what Morris was doing. Morris, after meeting over dinner with Lloyd Morgan, Crawford and Foster (the new head of University College's Education Department), was charged with writing a series of articles dealing with the university scheme for the Bristol Times and Mirror. Crawford, Ferrier and Foster would do the same for the Western Daily Press.

Taylor then asked Morris to met and dine with Michie, the lead writer for the Bristol Times and Mirror. Morris was here told that this paper had not supported the University College Bristol since they viewed the institution to be under the control of Arrowsmith. However, the Board of the paper decided to accept articles from Morris with a newspaper editorial preceding Morris's work. To do this, Morris actually bought his first typewriter and used it to prepare three articles. They appeared on the successive Mondays of April 30th, May 7th and May 14th and were all in a good position in the paper. The first article was introduced with the newspaper editorial and each article ran about a thousand words.

But before they appeared, Morris had written a long letter to the paper under the heading "Technical Education in Bristol". Though he stated there was a significant need for such instruction, he did note that trade classes in plastering and plumbing, while producing better plasterers and plumbers, did not provide true education. His views also noted that commercial classes such as in shorthand, typewriting, French, German and elementary math were of great value but were in no sense educational. Morris further observed that:

> "... We hear continually of German traders getting the better of 'Britishers' by methods which do not show superior energy, for in that we are always their equals, but in catering for the wants of the people they serve; or in making them believe that they do; in fact by the exercise of imagination. Nothing is more stimulating to the imagination than study, and study to be effective must be properly directed ..."

Comparison to Sheffield University

The university that was closest to what Morris and others strived for at Bristol was at Sheffield. To find out more about Sheffield, Morris wrote to its Vice-Chancellor Hicks and asked if he could visit. Hicks and Morris had previously exchanged letters on university development in July 1905 and Hicks had been associated with Ramsay in work aimed at securing grants for the university colleges while Ramsay was Principal at Bristol. Morris particularly wanted advice on the matter of the future arrangement with the M.V.T.C.

Morris went to Sheffield for this meeting on February 14th and received a tour of the university's buildings and spent time with Professor Arnold (Metallurgy) and Professor Ripper (Engineering). Ripper, in fact, told Morris that fifteen years earlier some of the leading manufacturers ridiculed the idea of applying chemistry to their work but now they all had [or were putting together] their own scientific staffs. An item of additional surprise to Morris was the value of building and endowment at only £200,000; but the city gave £8,000 a year for applied science and £5,000 for general purposes—the County Councils gave another £2,500 a year.

Before the university was founded in Sheffield there was already a university college, a medical school and a technical school. Hicks and Ripper were against the idea of bringing in the M.V.T.C. but Morris had to insist on at least including the section that trained students to London degree standards as politically necessary. They further advised getting in touch with the Labor members of the City Council so that lectures could be arranged in the surrounding areas. This may have been the inspiration for Morris's letter on 'higher education' that would appear in April. However, Morris knew that Ramsay had had little success with this same venture while in Bristol; hence Morris only arranged a lecture series right before he left Bristol.

External Committee

On February 19th a meeting of the senate was called to consider a resolution that Morris had already presented to an informal meeting of the members of the Faculty of Arts and Sciences that had met their general approval. It read:

"That the Senate of the University College, Bristol, is of the opinion that the foundation of a University in Bristol can only be accomplished at the hands of a Committee consisting of leading citizens of Bristol, of persons of influence in the West of England, and of representatives of existing institutions.

That should such a Committee be formed, and should the Council of the College think fit to advise the Governors to take powers to place the funds, lands, and buildings in its hands for incorporation in a University for Bristol, the senate wishes to assure the Council that it will have its cordial support and expresses the opinion that no guarantee of the integrity of the College, or of the interests of the staff, other than that afforded by public enquiry at the hands of experts, is either necessary or desirable."

This resolution was passed unanimously with small wording changes even though Lloyd Morgan objected strongly but he vehemently denied any intention to oppose the measure saying only that he tried to behave in an impartial manner. It was fascinating that the views of the senate had changed completely in four months.

Two further developments occurred less than a week after this meeting. First, on February 22nd, Lloyd Morgan told Morris that Lewis and Francis Fry had called a meeting of those individuals who had agreed to support the university movement and that they would be following the resolution that had been passed by the senate. Second, on February 24th, Ernest Cook told Morris that, in his opinion, the city of Bristol should not give any money to competing institutions and that there was some talk amongst city government at cutting off all grants until University College Bristol and the M.V.T.C. came to an understanding.

Cost of the University

Morris had been insistent on putting together a statement on what the university would cost. On March 4th, he was informed that Lloyd Morgan and Tom Williams had been appointed to draw up a statement as to the financial needs of the university and that he also be involved. Of course Morris realized that he would have to do the work since Williams [who was a lawyer in partnership with Morris's friend Robert Carpenter] was totally ignorant of the modern university situation. Williams would though be an invaluable critic and would represent the general public who would, of course, have to pay for this new university. Morris completed this work and gave his report to this committee on April 23rd.

On April 30th, Morris had a meeting with Lloyd Morgan and Williams where he noted, in rough terms, the cost of the university in its first year would be £21,000; professors were to be paid £500 a year according to Morris in this model. Morris also proposed what the general committee, when it was formed, would need to do. This committee would also be made aware of: the conditions that Haldane felt necessary for the charter of the university; that the college council and staff of University College Bristol would cooperate and that the Teacher's Association support the idea of a university in Bristol. The committee would also be given copies of Morris's pamphlet that provided general information for universities in Manchester, Liverpool, Leeds, Sheffield and Birmingham.

With this background in hand, the general committee first approached the governing bodies of University College and the Merchant Venturers Society to get them to cooperate in establishment of the university. Then it was suggested that representatives of the two institutions, acting in cooperation with the general committee, approach the Bristol City Council and ask that body to appoint a commission of enquiry into the present position of higher education in Bristol and the cost of the proposed university. Morris believed that members of this commission should not hold or have held teaching appointments in Bristol and that previous members of it need not be academics.

However, Morris realized that there was much, much more to do with the university movement but he had tired, to a degree, of doing so much work for it and at his own expense; he was after all only being paid £350 a year. He began thinking that he should go on with his research work at low temperature with the idea in mind that success in it would allow him to obtain a better chair position at another institution.

Visit by Ramsay

Ramsay visited Bristol on February 17th, 1906 to open the chemical laboratories and art school at the Clifton High School. Morris was present at this dedication where Ramsay said that school subjects should be educational and not just useful. When they returned to Morris's laboratory [where Ramsay was both pleased and astonished to see so much research going on], Ramsay reopened the Indian question again. He urged Morris to allow his name to be put before a committee of the Royal Society that would meet to recommend to the Secretary of State for India a candidate for the post of Director of the new Institute. Morris gave no reply at that time but wrote to him, after significant reflection, on February 19th where he said in a letter:

"I have been considering the pros and cons, and have come to the conclusion that, as things stand at the moment, I am prepared to let my name go before the committee.

I have already refused to consider the matter on more then one occasion for the reason that I wished to devote myself to research work, and teaching, and to keep clear of administrative work as far as possible.

Circumstances have, however, driven me into the university problem at the expense of research work; and under conditions which are not at all to my interest. Here I am carrying on administrative work, which lies outside my province, without the help, to say nothing of the opposition of those whose duty I am doing, or who should at least have given me their support. The success of my policy depends entirely on the influence I can exert; and my work is accepted by the authorities, who at the same time refuse me the recognition which would render it more effective through fear of hurting the feelings of inefficient seniors.

I have not the least fear that the matter will not go through, in spite of the lack of knowledge on the part of the Council, and timidity on the part of the Principal. I believe too that I am not the only holder of this view. Cook, afterward Sir Ernest, Chairman of the City Council Education Committee, told a friend of mine the other day that 'Travers is the only man connected with the College who knows what he wants, and that even his enthusiasm could not move the mass of inertia opposed to him'.

At the same time I am not acting fairly to myself in the matter. I have given up all outside work, not to speak of my research work, and am put to heavy expense in making enquiries, visiting other universities, and even entertaining in a small way. I make, when my subscriptions are paid, less then £300 a year.

It seems that whether I go to India or stay here, I shall have to undertake administrative work, and in India from a much higher position, and with greater advantage to myself.

I am not dissatisfied with what I have done here. I have studied the university problem by reading all the matter I could lay my hands on, and have visited all other English universities except Birmingham. I have established the principles of the constitution of a modern university in the minds of the Council . . . I have obtained the opinion of the government on the Bristol university question, and have given the Council the lead in obtaining the support of men of influence in the West of England. I have obtained for the College the support of the Conservative press, and taken entire charge of College advertisements (press notices). In internal matters there is little, from school scholarship schemes to the organisation of extension lectures, which I shall not leave my mark on, and every point has been gained by fighting, often against mere prejudice . . ."

Thus all of Morris's frustration and feelings about Bristol were so expressed. Ramsay wrote to Morris on March 2nd stating he would probably be offered the position in India. Morris, later that same day, told Lloyd Morgan and Lewis Fry about this and they were both complimentary to Morris and said they were sorry that he was leaving. Soon afterward all in town and at the college knew and many asked Morris if it was true that he was leaving. On March 5th, Morris wrote back to Ramsay saying that he would accept the position if it were offered to him.

But it was not as simple as Ramsay had thought. The committee consisting of Ramsay, Glazebrook, Starling, Tilden, H. Brown, Martin and Prain formally met and Prain suggested that Wyndham Dunstan [chemist and director of the Imperial Institute] should be considered. This seemed to end Morris's candidacy. Morris, realizing this was out of his hands put the matter from his mind though he did write to Lewis Fry and update him on this development. Fry replied saying—

" . . . I shall be much interested in hearing the result of the next R.S. Committee. I wish you success in your candidature, but I shall regard your appointment with mixed feelings, as you know."

End of Spring Term and Social Life

The last event of the 1906 spring term was the March 31st dinner of the University College Colston Society. Morris's pamphlet was distributed to the guests and advance copies sent to the press and a notice of it had appeared in the Bristol Times and Mirror on March 10th. The pamphlet also received notice in a long leader on the Colston Dinner, which appeared on the morning of April 1st. The President of the Society, Arrowsmith, announced that four gentlemen had promised Mr. Lewis Fry to contribute the amount of £30,000 to acquire the Blind Asylum site. Of course no mention was made of Morris's name in connection with this. Arrowsmith also spoke of raising £150,000 to £200,000 for the university; this greatly amused Morris given the recent *severe* financial difficulties of the university college.

During all this Morris kept little record of his social life though he did have many friends. Morris and Ferrier shared a house; Ferrier had a study and dining room on the ground floor and Morris had his study and dining room on the first floor. Their bedrooms and a spare room were above. They dined as each other's guest in alternate weeks.

Ferrier kept late hours whereas Morris did not. Morris, retaining traces of his illness of a few years ago, had to be careful about his health. He still had stomach troubles on and off, possessed an occasional stammer at times and a cramp in his hand that would come and go. He also would be tongue-tied at times and could not start a sentence. His signature was also usually illegible and he was a poor enough sleeper that his bedroom could not face a road.

Even with these health issues, Morris and Ferrier entertained often and jointly. Morris almost always handled the arrangements for these events. Sometimes ladies were amongst their guests; but they were usually only a party of four, with others coming in after dinner to make up two tables for whist. Ferrier was a master at this game and used his extraordinary memory [he could lecture without notes] to advantage while playing. Morris, unfortunately, was not very good at whist.

Efforts at Research

When Morris arrived in Bristol he found an air liquefier in the basement that neither Young nor Morris could get to work. Ultimately, Morris rebuilt the compressor with the aid of Ferrier's mechanic and bought a Hampson air liquefier.

It was Morris's intention to try to liquefy helium and as a first step he proposed to determine its critical and boiling points by a method similar to that which had been used by Olszewski to find these constants for hydrogen. Since the results of temperature measurements made by electrical methods were meaningless [as noted by Olszewski for hydrogen], Morris thought that he could measure the temperature of the helium by means of a constant-volume helium thermometer consisting of a small steel bulb with a stem made from fine (hypodermic needle) steel tubing, leading to a pressure gauge.

The thermometer was enclosed in a small steel vessel and surrounded by three paper-thin shells of glass; 1/2 mm steel tubing connected it with a compression pump and a gasholder. It was proposed to cool the containing vessel in solid hydrogen, fill it with helium at various pressures, and after expanding to various low pressures to then note the temperature.

A small steel hand-operated compression pump was made. Morris also had a gasholder made from steel plates that were welded together with air/acetylene. Morris had recently arranged for the first exhibition of this process in the west of England and had converted Ferrier to its use by repairing one of his favorite golf clubs! The gasholder was made with a narrow annular well to hold mercury, in which the dome would float, thus avoiding the use of water.

Unfortunately, Morris's work for the university scheme prevented him from ever putting this apparatus together.

University Doings—Late Spring and summer—1906

On April 19th, Morris received a circular letter from Fry asking that he join the general committee for the university movement. Morris accepted the invitation and on the 23rd told Fry

> " . . . that, supposing I were staying in Bristol I was willing to act as honorary secretary to the committee, but I would not accept remuneration from the Colston Society, giving reasons why I would not."

In the course of the next few days, Morris spoke frequently with Fry who was quite satisfied with his progress in interesting important people in the university scheme.

A preliminary meeting of this general committee was held on June 1st. Morris's notes on it were as follows:

> "Present—Lewis Fry, Lloyd Morgan, Arrowsmith, T. Williams, and myself. I had prepared a statement under three headings, (1) The needs for the establishment of the University, (2) The cost of the University, (3) Means to be taken to secure our ends, involving the appointment of the Commission of Enquiry. T.W. and A. were very despondent about the whole affair, and inclined to take the line that the time was inopportune—Was it ever opportune for such an appeal? L.F. behaved splendidly; if he does not feel sure of success, he has at least the courage to go on, and make a fight for it. If it had not been for him the whole scheme would have been dropped.

As one with no authority, little influence, but a keen interest that the
matter should go forward, I had an anxious time. L.M. didn't count.
Finally L.M. and I were instructed to draft a statement for the General
Committee."

In early June Morris went away fishing for a few days with Francis, Carpenter and others. Upon his return on June 6th, Lloyd Morgan gave him a draft of their memo for the general committee, which he spent some time rewriting and putting its facts in order. The next day he went through this memo with Lloyd Morgan and sent it off to be printed.

On June 9th, Morris wrote to Ramsay telling him that they had just received £10,000 from H.O. Wills. This amount, combined with the £30,000 in pledges for the Blind Asylum Site and £1,500 from a man named Whittuck, made a good sum to go to the university committee with.

A second preliminary meeting of the general committee was on June 12th. Present at this meeting were Lewis Fry, Lloyd Morgan, T. Williams, Arrowsmith, George A. Wills, Napier Abbot and Morris. Abbot was a lawyer, a brilliant and scholarly man who was though a pessimist by nature. Lewis Fry was chosen as chair of the committee and Wills was chosen as its honorary treasurer. When the question of who would be secretary arose, it was suggested that Morris be the honorary secretary but Arrowsmith instead wanted his clerk Davis in this role with Morris as the honorary organizing secretary. No one other than Morris saw that Arrowsmith was trying to control the university movement. After Morris objected to Arrowsmith's maneuvering, it was decided that Morris be honorary secretary and that the school's registrar Rafter assist him. This decision led to Arrowsmith being very withdrawn the rest of the meeting and taking no part in the discussion on the sum of money to be asked for; he even left the meeting early.

Immediately after the meeting, Morris wrote to Ramsay and said that the university business was going through a critical phase and that two-thirds of the general committee favored giving up the whole affair. Morris though vowed to Ramsay that this would not happen!

The battle for the position of honorary secretary was not over yet. Early the next morning Lloyd Morgan told Morris, who was on his way to lecture, that Arrowsmith had complained, in writing, to Fry that his suggestion about Davis had not been followed. Morris told Lloyd Morgan that they must stick to what had been agreed to.

After his lecture, Morris was asked to see Lloyd Morgan who told him that Fry had come looking for him during his lecture. Fry had spoken about Arrowsmith with Lloyd Morgan who made the mistake of saying that Morris would fall in line with the views of Arrowsmith about the honorary secretary position. Morris, very angry, told Lloyd Morgan that he had no right to speak for him and that only if the general committee formally asked him to step down would he do so. Fry, who had in the meantime gone to meet with Abbot and Williams, later came back to the college to see Morris.

Fry, very upset, asked that Morris agree with Arrowsmith's request. Morris replied—"I would do almost anything you asked me to Mr. Fry, but I will not wreck the ship." Morris proceeded to go over past history, mentioning the advice given by Haldane, himself a Liberal, that the committee should be mainly Conservative. Arrowsmith's name was anathema to a large number of people that they wanted support from, particularly to the members of the Merchant Venturers Society. Morris also mentioned Taylor's letter where he had referred to Arrowsmith as an obstacle to the support of the college by the Conservative press.

Morris then saw Williams the next morning. Williams was not pleased with this meeting; after this meeting Morris again met with Fry and he again told him that he would not retreat from his position and also noted that Arrowsmith would not resign from the committee since he had apparently said all that he would. To this Fry simply shook his head.

Soon afterwards, Fry had apparently soothed over matters with Arrowsmith. At the Clifton College garden party a few days later, Fry told Morris that he—" . . . had got Arrowsmith to meet the Bishop of Hereford at his house, and all was well." However, this was not the end of problems. The next day [Sunday] at supper at Carpenter's, Hall told Morris that he had met Williams at the Liberal Club during the week and had asked him " . . . how the Travers University was going on." Williams answered—" . . . that he had just touched the spot and unless Travers consented to behave more reasonably there would be some resignation from the Committee." Hall pressed for details but only found out that it had nothing to do with the appointment of Davis as secretary. Fortunately, no one resigned from the committee and this was probably due to Arrowsmith who dearly loved the college even if he had a massive ego, which compelled him to control everything associated with the college.

Arrowsmith did have one more little dig at Morris though. He wrote to Fry saying that there should be a resolution compelling the honorary secretary to keep the committee informed of all that was occurring. Fry showed this to a laughing Morris who though thought it unusual to open proceedings of the committee with a vote of censure on one of its officers.

The general committee next met on July 2nd at the Chamber of Commerce. One hundred invitations had been sent out; forty were present and thirteen had written and regretted their absence. Arrowsmith arrived at the last moment and Morris, seeing him at the door, went over to him, shook hands, and said that he had kept a place for him next to himself. Arrowsmith sat down and was heartened after Fry's opening speech praised him. The general committee was then formally constituted with George A. Wills as honorary treasurer and Morris as honorary secretary. An executive committee was then put together and began to consider the financial aspects of the problem.

In the course of the next week, Morris had a long talk with Napier Abbot to whom he had to make it clear that though he wanted to get out a scheme as soon as possible, he did not wish to rush things and that the scheme could be viewed as a beginning. Morris also had long talks with W.W. Jose and T. Butler who were leading members of the Bristol City Council. Butler in fact thought

> " . . . that we shall get the money out of the City. The Council cannot say
> no:—He is going to talk to members of the City Council for me."

The next few days were quiet and on July 13th Morris went to visit his family in London before leaving for Norway and fishing with his friend James Chick. Morris had heard that the Indian post would probably be offered to him so he asked his brother Harold to open letters forwarded to him from Bristol and cable him if anything on India turned up. Fishing in Norway with Chick was relatively poor and Morris returned to London where he found a letter from the Secretary of State for India formally offering him the Directorship of the Indian Institute of Science, which he accepted.

Fall 1906 and afterward for the University

Immediately, Morris resigned his chair within the college; Francis was soon appointed to this position. Morris and Francis exchanged regular visits when Morris was in England and Francis remained Morris's closest friend; he would be Morris's best man at his wedding and godfather to his son.

Morris also gave up his secretary position on the general committee for the university. He was asked to attend an informal meeting to choose his successor on October 5th; this meeting had Lewis Fry, Arrowsmith, Abbot, Williams and Lloyd Morgan also present. Arrowsmith wanted Rafter, the registrar, to act as secretary, which would have been disastrous for the university. Morris campaigned for his colleague Professor Rowl (English Department) to take on the work because he understood the problems of the modern university. Rowl was nominated after an hour and a half meeting with Rafter to act as clerk.

On October 9th, Morris attended a formal meeting of the general committee and since Lewis Fry was ill, Arrowsmith took over and moved a very appreciative vote of thanks to Morris. Cowl and Rafter were also elected to their positions.

Morris's last act, with respect to the university movement, was to contribute 100 guineas to the university fund; this money was to be used to pay, in part, for an assistant for Cowl. Morris felt, in telling Fry, that no man carry the load that he had, in addition to his regular work, during the past two years. It turned out though that the money was not used for this noble purpose.

Cowl's efforts for the development of the university were excellent but they made him many enemies [one report had Cowl neglecting students in the junior class]. Ultimately, after University College Bristol and the M.V.T.C. had started to become the University of Bristol in 1909, Cowl's chair was advertised and he was not elected to it. He left Bristol in 1911 threatening legal action against Lloyd Morgan for slander. Lloyd Morgan told Morris in 1912, when they met at the Congress of the Universities of the Empire in London, that the university had been in the wrong and that the reason they would not reinstate Cowl was financial. But, in the meantime a man's life had been ruined.

Bristol honored Morris not only with letters of congratulation but also on October 12th at a public dinner in his honor presided over by Lewis Fry. He reviewed the position of the university movement, saying with regard to Morris:

> "Dr. Travers had written an admirable pamphlet on the university problem and the need for a University in Bristol, and had used his influence among his friends and acquaintances in pushing forward the idea of a University for Bristol. His cheery optimism had helped the movement in a way that he could not too fully praise. He had given an impetus to the university movement more than perhaps any other person . . ."

Fry then next referred to the acquisition of the Blind Asylum site and also to the formation of the general committee, of which Morris had been secretary.

Morris, when speaking, developed the idea that the Bristol University should not be just a copy of one of the universities established in other great cities, but should be peculiarly suited to the needs of Bristol and the west of England. The citizens of Bristol had to create their own university, which Morris typified with " . . . a Colston department of economics and a Cabot department of economic geography." Morris enjoyed the evening immensely and felt very proud of all he had done.

Morris left Bristol in 1906 with many friends and no enemies. He had almost literally fought with Arrowsmith on several occasions; but the two men developed more than mutual esteem, even affection. In 1909, when Arrowsmith heard that Morris was to be married he sent a wonderful present and asked that Morris bring his wife to see him if Morris ever came back to Bristol. Morris and Dorothy came to see Arrowsmith one Sunday afternoon and had tea; after this Morris never saw Arrowsmith again.

Morris had the warmest regard and affection for Lewis Fry and every time that he visited Bristol, until Fry's death at eighty-nine in 1921, he called on him. Even at the end of Fry's life, when his butler said he was seeing no visitors, he still asked Morris in for a few minutes to talk. It is to him that Bristol owes its university.

Events in 1907 with respect to the university saw exchange of communications with the Merchant Venturers who were unwilling to agree to a genuine amalgamation of the two entities into a university. Later in that year, negotiations were suspended until a contribution of £100,000 could be made for the endowment of the university. This large contribution occurred in January 1908 when the President of the Colston

Society George A. Wills announced the promise of £100,000 for this endowment by his father H.O. Wills.

Letters from Fry on March 21st, 1909 and Francis on April 11th spoke of further success. With £200,000 in the hands of the general committee, and the assured backing of the City Council—largely due to Dr. Cook's efforts, a plan had been submitted to the Privy Council, under which the two entities [University College Bristol and the M.V.T.C.] should be absorbed into a university under certain conditions.

It was proposed that there should be a Merchant Venturers Faculty of Engineering, which would also include the Engineering Department of University College Bristol. This body should include a Department of Applied Chemistry, in which the M.V.T.C.'s Principal Wertheimer would be a Professor. This faculty would only teach the chemistry, physics and mathematics that were needed by the engineering students. All other teaching in these subjects should be transferred to the university college.

The Privy Council then expressed official approval of the general committee's scheme and when the final opposition of the Merchant Venturers collapsed an agreement was reached. This led to a decision to enlarge the chemical laboratory and a final agreement with the Merchant Venturers. Following that, the City Council was approached for a definite grant and then the hope was that a charter would be granted immediately afterward. This charter was granted on May 24th, 1909 when Edward VII affixed his signature and the Royal Seal to the charter.

In October 1909 the University of Bristol opened its doors to 288 undergraduates and 400 other students with women welcomed on even terms with men and H.O. Wills III as its first Chancellor. Today, in 2011, it has over 23,000 students spread out over 34 departments and 15 research centers with six separate faculties. It is ranked as one of the top ten universities in the United Kingdom and has received more applications per place than any other British university.

CHAPTER IX

Development and Founding of the Indian Institute of Science

Development of the Idea

As the 19th century drew to a close the only universities in India were: Calcutta [Kolkata as of 2001]; Bombay [Mumbai as of 1995]; Madras [founded in 1857]; Punjab at Lahore [founded in 1882] and the University of Allahabad [founded in 1887]. However, these institutions functioned only as examining bodies while the teaching was carried out in numerous colleges. Consequently, educational standards suffered significantly.

In 1889 while giving a convocation address, Lord Reay [the Governor of Bombay and Chancellor of Bombay University] argued for a teaching university that would be able to attract talented people from the West by stating:

> "It is only by the combined efforts of the wisest men in England, of the wisest men in India, that we can hope to establish in this old home of learning, real universities which will give a fresh impulse to learning, to research, to criticism, which will inspire reverence and impart strength and self-reliance to future generations of our and your countrymen."

Following up on this, in the early 1890's a wealthy Parsee merchant and mill owner of Bombay by the name of Jamsetji Nusserwanji Tata (J.N. Tata), who had traveled extensively in Europe and America and is widely thought of in 2011 as the "father of Indian industry", came to the realization that the industrial progress in Europe and America had run parallel with

the development of scientific and industrial research for which facilities were available in universities and technical institutions.

Before a definite plan could be proposed and implemented, Tata, realizing the grave need for highly educated Indians, offered assistance, in the form of loans, to distinguished graduates from any of the five Indian universities who wished to go to England for further study. This was called the Tata Education Scheme and though it perhaps was capable of improvement in detail and enlargement in its scope, it did though offer the possibility to young Indian men of devoting themselves to a life of study and research rather than one of earning a livelihood.

In November 1896, Tata wrote a letter to Lord Reay saying—

> "The improvement of University Education is the key of all educational improvement,"

and suggested the creation of a national university.

To learn about scientific institutions and universities in a variety of countries, Tata sent his ward and close aide Burjorji Jamaspji Padshah to visit and learn about such institutions in England, continental Europe and America. Padshah spent eighteen months doing this research and after returning, prepared an outline for an "Institute of Scientific Research for India". Tata and Padshah took John Hopkins University in Baltimore in America as their model, in part, since it had the distinction of being the first institution in the world to be founded as a post-graduate entity.

To begin developing such an institution in India, Tata set aside properties in Bombay in October 1898 so that they would generate Rs.125, 000 or about £8,500 for the endowment of a university institution for post-graduate study and research. This small amount would hopefully inspire others to give for this institution.

The next step in 1898 was to invite a number of the distinguished citizens of Bombay to form a provisional committee that included: Lord Reay; Bombay University Vice-Chancellor E. T. Candy (the well known oriental scholar and Judge of the High Court) as Chairman; Justice Mahadev Govind Ranade (who was a well-known Indian scholar) as Vice-Chairman; Tata and Padshah as honorary secretary. Approximately twenty others were also on this body.

This committee met and drew up a report, dated December 13th, 1898 that became public on December 31st, 1898 when it was presented to Lord George Curzon, the new Viceroy of India. This document was headed—"A Research Institute for India". It proposed three departments or faculties as they would be called when more fully developed and they were:

i.   A Scientific and Technical Department
ii. A Medical Department
iii. A Philosophical and Educational Department.

The first Department included:

a. Physics and Mathematics—Advanced courses in all departments, including mathematical physics and electrical engineering.
    Staff—One Professor and one Assistant Professor.
b. Chemistry—Advanced inorganic chemistry, organic chemistry, analytical chemistry and agricultural chemistry.
    Staff—Two senior professors, two junior professors.
c. Technical chemistry applied to different arts and industries.
    Staff—One professor.

They were to be have the following facilities:

A physics laboratory,
A chemistry laboratory with departments,
A technological museum and
A library.

The medical and the philosophical departments were to be organized similarly.

The Viceroy told the delegation presenting him with the report that though he considered their work carefully he could not give them a final answer but the proposal had his warm sympathy. He raised three points: was the committee satisfied that after they selected their professors there would be a sufficient number of advanced students for them to teach; would these students, after their training, receive employment and that he had misgivings about the proposed philosophical and educational department—he saw, without actually saying so, the possibilities of this department developing along political lines. In conversation occurring after the Viceroy raised these concerns, it was stated that the Dewan of Mysore had a fund of Rs.550,000 available for this plan.

The committee's report, together with an account of the presentation to the Viceroy, was printed and copies were widely circulated. Opinions on the scheme were received from some seventy individuals, most of whom were government officials. One comment was that putting forward proposals involving large sums of money might actually do more harm than good; a beginning might be made though on an experimental basis and on a small scale. A second comment involved the philosophical and educational department; most seemed against this entity.

The scheme, and the comments about it, was then sent to the government of India who found that one item it contained was unacceptable. Tata proposed to "transfer" the whole of his real estate holdings in Bombay to a trust, which should take them over and manage them, paying one half of the income generated to the research university and one half to Tata during his lifetime and after his death to his male heir in perpetuity. Subject to the withdrawal of this condition by Tata, the government [which feared it being party to perpetual litigation if the condition remained] recommended that this proposal should be discussed with the members of a conference meeting at Simla.

Dropping this condition did not happen immediately though. Tata discussed the scheme and this condition with the Secretary of State for India, a Lord George Hamilton who supported it! Ultimately, Tata decided to drop his condition when it became apparent that both of his sons were likely to be childless; at that point he really lost interest in the family settlement.

The Simla conference included: Thomas Raleigh, member of the Viceroy's Council—Home Department; J.N. Tata and his aide Padshah; Justice M.G. Ranade; the Surgeon General for India and four senior officers of the Indian Education Service. The conference drew up a report [dated December 18th, 1899] titled "The Indian University of Research" in deference to the wishes of Tata and advised that departments i and ii should go ahead but *not* so for department iii. It also advised that, since it had little if any knowledge of modern developments in universities or technical institutions, a European expert on university education should be invited to visit India to examine and advise on the scheme. The scheme was also welcomed in the report but no commitment was made by the government to support it financially or otherwise.

Ramsay's Visit and Plan

At this point, Tata invited William Ramsay to visit during India's cold weather season of 1900-01 and give him advice on the scheme. Ramsay was given a copy of the 1898 provisional committee report and was told that approximately Rs.700,000 (£50,000) was available for capital expenditure and Rs.232,000 (£15,000) was available in donations. He knew from the outset that finances would not permit anything more than two or three of the subjects in department "i" from being started in such an institution as Tata envisioned.

After reviewing this information while in Bombay, Ramsay made short visits to Poona and Bangalore. Before coming to Madras Ramsay had formed the opinion that the standard of scientific and technical work in India was very poor, though he thought highly of the work that some individuals were doing.

In Madras he saw a new venture developing—working with aluminum that was due to establishment of the Indian Aluminum Company, Ltd. A similar project for the chrome tanning of leather was in development. These industries, new to India, deeply impressed Ramsay and made him consider embodying this "spirit" in what he would propose to J.N. Tata's committee. Ramsay even confided to Morris that the man who held the office of Director of Industries would be a good fit as the first Director of the Indian University of Research.

Ramsay suggested beginning the Institute with Departments of General Chemistry, Engineering Technology, Electrical Technology and Industrial Technology.

There should be also Assistant Professors who should be specialists in different branches of industry and who should engage individually in initiating industrial processes. As the development of any industrial process progressed, the assistant professor in charge of it would become its Managing Director and the students who had helped the professor to develop that specific industry would staff the process.

Both the provisional committee and the government of India were disappointed with Ramsay's report to them. The government in particular believed that this plan would mean that the research university would be embarking on financial risks that the government would be responsible for. As a result, they rejected Ramsay's proposal as impractical and suggested that David Orme Masson [Professor of Chemistry at the University of Melbourne], whom Ramsay had proposed as the entity's first Director, should visit India. He should work with Lieutenant Colonel John Clibborn, Principal of the Rurkee Engineering College, to develop an "acceptable scheme" for the research university.

In their report, dated December 1901, Masson and Clibborn outlined a scheme for an institution which should start as a "school of experimental science" consisting of departments of chemistry, experimental physics and biology given that funding was quite limited. They also proposed the name "The Indian Institute of Science" and they stated:

> "The Institute should in the first place train students and turn them out imbued with the spirit, and practised in the methods, of experimental investigations, thus contributing directly to the intellectual life of India, and at the same time aiding its industrial progress. In the second place, it should, by the work of the professorial staff and senior students, steadily carry out and publish original researches of scientific interest and, it may be, of practical value; and then researches should, as general rule, deal with materials and problems of special importance to India. In the third place the Institute should become in time the accepted central authority on any scientific question within its own domain, and should in this capacity deal with, and report upon, problems referred to it by the Government of India."

Their report included a detailed estimate of current expenditure of £12,000 and an approximate estimate of capital expenditure of £45,000 for the Institute.

Advanced University Education in India

A major difficulty with developing the Institute was that virtually no one in India knew anything about advanced university education. The members of the education service were mostly university graduates whose knowledge and experience with a university was what they acquired while undergraduates in British universities. Universities in India were examining bodies much like in Britain and when Morris came to India in 1906, he found that the regulations for the B.Sc. degree in the Bombay University were *identical* to those seen for London in 1860, which is the year that the degree in science was first introduced.

The Viceroy also recognized this sorry state of affairs and established the office of Director General of Education; but the holder of this office was a specialist in primary education and as a result nothing was done to further the cause of university education. In fact, Morris noted that he and this Director were in India from 1906-1914 but never met and that this individual never had his name on any of the papers related to the Institute.

In addition, the Viceroy appointed a royal commission on university education in India and when he received their report, he sent for the commissioners, ripped up the report in front of them and stamped up and down on top of its shredded pages. Morris later read a copy of this report and he understood why the Viceroy did what he did.

Efforts of Charles Martin

Following the visits and reports of Ramsay and Masson/Clibborn, Tata asked Charles Martin [Professor of Physiology at Melbourne University in Australia], who had spent a good deal of time in Bombay advising the government in connection with the production and use of plague serum, for advice on his idea. Tata and his associates could have found no one more qualified to advise them than Martin. In looking over the aspects of the proposed Institute, Martin soundly advocated for something very small and practical at the outset; thus implying that Padshah's research university was just a dream.

Location of the Institute

J.N. Tata had established a modern silk farm near Bangalore in the 1890's and had become acquainted with the Dewan of Mysore, a Sir Seshadri Iyer [regarded as the "maker of modern Bangalore"]. At a later time, with the idea of an Institute for research beginning to move forward, Iyer approached Regent Maharani Kempa Nanjammani Vani Vilasa Sannidhana (ruling for her minor son Maharaja Sri Krishnaraja Wodeyar Bahadur IV) and asked for support for the Institute, which was readily given.

In time more than 371 acres of land were provided for the Institute thus ensuring that it would be in Bangalore. This site also had the advantage that its climate was temperate nearly all year; it was not too hot for Europeans nor too cold for the natives. It was also accessible from Bombay and Madras and within reach of Chota-Nagpur, a district rich in mineral resources.

The Tata "Sons"

J.N. Tata had two sons, Dorabji and Ratan and he was, as previously noted, the ward for Padshah.

Dorabji, the elder son by twelve years and born in 1859, was educated in England and continued his studies at St. Xavier's College in Bombay where he received a degree in 1882. He worked for two years as a journalist for the Bombay Gazette and then learned the cotton trade in Nagpur. Dorabji was instrumental in the establishment of the Tata Steel and Tata Power Companies, which are the core of today's Tata group. He also was extremely fond of sports and was a pioneer in the Indian Olympic movement.

Ratan Tata, the younger brother, was also educated at St. Xavier's College and looked after the Indian affairs of the L' Union Fire Insurance Company and was in charge of the trading firm Tata & Co. which had offices around the world and traded in cotton, yarn, silk, pearls and rice. His real interests though lay in social and philanthropic causes and he regularly donated large sums of money to combat poverty and destitution in India.

Burjori Jamaspji Padshah was born in 1864 and came from a highly talented family. His father died prematurely and left Padshah at 16 to run the family business though he was fortunate that his father's best friend was J.N. Tata who made Padshah his ward. Padshah was a man talented in many areas, including: physics; political economy and mathematics though he never took a degree at either of the universities he studied at—Bombay or Cambridge. As an educator, scholar and thinker, he had few equals and there was no subject that he did not know and could not discuss in an intelligent manner. He never cared for reward or appreciation and ceaselessly did all he could for the advancement of his country and the good of humanity [he, in fact, had a close personal relationship with Mahatama Gandhi].

However, even though his sons [and Padshah] were remarkable men in their own right, J.N. Tata did not involve them in his business or with his public activities.

Death of J.N. Tata and his Will

J.N. Tata again became ill in January 1904 and he went to Europe for rest and treatment. Before he could do this, his wife, Hirabai Daboo Tata, died in March 1904. Soon after her passing, he went to Bad Nauheim, Germany where he died of heart disease on May 19th, 1904. He was subsequently buried in a cemetery near London.

Upon hearing of Tata's death, the government of India wrote on June 1st, 1904 to the government of Bombay expressing their condolences and wondering what might happen to the plan for the Indian Institute of Science. They replied that

"Mr. Tata's younger son, Mr. R.J. Tata, had written, on behalf of himself and of his elder brother, Mr. D.J. Tata. Offering to transfer the properties scheduled for the endowment of the Institute to the Treasurer of Charitable endowments".

The opinion of the Advocate General for Bombay was:

" . . . that Mr. J.N. Tata's sons, as Executors, and Sole Residuary Legatees, had power to carry out his scheme, and generally to carry out the terms of the award on the valuation of the properties, which had been completed and accepted, and which represented an agreement between Government and the Testator, the terms of which were binding on the Executors".

The operative clauses referring to the Indian Institute of Science in Tata's will, dated December 16th, 1896, were:

(11) "I also declare that I have recently formed a project for the formation of an indigenous University in this City [Bombay] upon a broad basis as Institutions of a like nature which have been founded through the munificence of private gentlemen in Europe and America to the intent that such University may become the means, along with several others of meeting the growing scientific needs of this country, and I intend to make the following provision for the furtherance of the object namely . . .

(12) I authorise my Executors to do all matters and things which in their judgment appear necessary and expedient in connection with the carrying out of thesaid project . . ."

It was quite curious that the Indian government did not demand and get an actual physical copy of the will of Tata at this time. Interestingly enough, soon after the death of their father, Dorabji and Ratan, offered to transfer the endowment properties to a trust on terms agreeable to them.

Believing now that all was well and would proceed smoothly, the Indian government and the Dewan of Mysore agreed to increase their subsidies, so as to bring the fund available for capital expenditures to Rs.750,000 and the income to Rs.260,000.

CHAPTER X

Setting up the Institute

Appointment of Morris Travers

Morris was formally offered the position as Director of the Indian Institute of Science by the India Office—London in a letter dated August 16th, 1906. The position would have a salary of £1800 per year and would have a pension of £500 per year after ten years service and a pension of £750 per year after fifteen years of service. It would not though be considered a government employee.

Morris accepted the position and asked when he should be expected to leave for India; he found it was to be some time in October. Ramsay was pleased to hear Morris's good news and he suggested that Morris call on the Tata brothers and contact Padshah; he gave Morris their addresses in London.

He immediately wrote to Dorabji and received the following reply of September 18th, 1906:

> "Mr. Padshah keeps me fully informed about what is being done in connection with the research University, and of course I had early intimation of your appointment to the post of Director. I am very glad indeed that you have at last been persuaded to accept this post. It was what we had been waiting for for a very long time, and Sir William Ramsay will tell you that even as late as last year I asked him to make another effort to enlist your sympathies with the aims and objects of the Institute my father wished to see established for the good of India. So much depends upon the start and the man who makes it, that now I am very glad that we have had so many years of weary waiting as it has enabled us to secure your co-operation, which perhaps might not have been the case if we had started, say three or four years earlier . . . I am looking forward to meeting you before you leave for India . . .".

Morris put this letter along with his other papers of little importance but after he met Dorabji in Bombay in July 1907 he realized how important it was. Had he known in 1906 of the name "Padshah" and the term "Research University", Morris would have resigned the Directorship immediately.

As recommended by Ramsay, Morris called on both Ratan and Dorabji. While Ratan had little, if any interest in the Institute, Morris noted to Dorabji that he did not consider Ramsay's proposals workable on the starting of industries. Dorabji was disappointed in hearing this and Morris wondered what he actually knew of the history of the plan.

Dr. Charles Martin and the Institute

Shortly after his appointment, Dr. Charles Martin, the Director of the Lister Institute, wrote Morris and suggested that the two of them meet Padshah for dinner. Morris thought that Padshah was not interested in the same type of Institute as Morris was which was an institution providing facilities for advanced study and research in science and technology. Padshah spoke of the importance of having Indians aware of their historic past and wanted Morris to meet people interested in Indian art, ethnology and archaeology. Morris, looking to the future, was not excited with this idea but did, in Padshah's company, make some interesting visits to Cambridge and elsewhere.

Martin told Morris that J.N. Tata did not want his name associated with the Institute, even in the naming of a chair or a library. Tata thought that when a founder's name was attached to an institution, other possible benefactors would not donate to it much as had occurred with Mason College in Birmingham.

Martin also said—

> "One of the functions of an institution like yours is to accelerate the establishment of self-government in India."

Morris also received from Martin a very complete collection of reports and documents related to Tata's plan. This helped bring him somewhat up to date with the scheme though it took more time for him to fully understand all of its details.

Further information about the Scheme and India

However, Morris could get no information about the Institute from the officials in the India Office. The matter was not in their hands and they certainly did not want it to return to them. They felt it would have come back to them if Morris knew about the relationship of the Tata brothers and the government and also the feelings of the India Office toward Padshah.

Before he left for India, Morris talked to many people who he thought could help him learn about the state of higher education in India and the Tata plan. After these conversations, Morris felt he knew less and would not learn anything until he arrived in India.

Morris left London on November 1st, 1906 for the first of his eight voyages between Europe and India. Arriving on November 15th and while waiting to disembark, he was asked by fellow passengers if he had seen the Times of India newspaper. Thinking that it must contain some news of importance, Morris picked up a copy, and to his astonishment saw, a picture of himself on the front page! Policemen in Bombay over the next few days duly saluted him.

Seeing India—Bombay and Poona

Morris had dinner his first evening in India with Sir Lawrence Jenkins (the Chief Justice of Bombay and later the Chief Justice of India) and Herbert Risley (Secretary to the Home Department—Government of India) who had handled the Tata case since 1902. Risley, who had conducted the last Indian census and was known as an ethnologist and was writing a book about India's people, however was a sick and bitter man who knew that he would be passed over for membership on the Viceroy's Council.

When dinner was over, Risley asked Morris what he would do with the Institute; Morris said that he could not answer until he knew more about the quality of the students. He first wanted to see as much as he could of India's colleges and he proposed to tour a number of them. Risley, who had a very poor opinion of Indian students, agreed that a tour was a good idea. The two men also talked about the modern university movement in England; Risley knew little about this but realized quickly that Morris knew a lot about it. At the end of their conversation, Risley said that the government wanted to help Morris as much as it could and also warned Morris—"If you play with politics we shall be very hard on you." This statement meant nothing to Morris at this time but would later prove to be prophetic!

Morris thus visited several institutions in Bombay and observed that teachers in the Bombay colleges were similar to the masters in English public schools and that they were able, hard working and conscientious but struggling to make good under impossible circumstances. It was possible for a student to obtain the B.Sc. degree in chemistry by memorizing what was in a few textbooks and by taking a course in elementary qualitative analysis. While the standard of the papers for the M.A. degree in chemistry appeared high, practical training for this degree did not go past elementary quantitative analysis.

Research work in Bombay was almost entirely confined to the Plague Investigation Laboratory at Parel, where the staff was mainly members of the Indian Medical Service. Outside of this facility, Morris met only two Indian researchers in Bombay and Poona. One was a Professor Gaggar who worked in chemistry. A second individual was Naegamvala who as Head of the Observatory at Poona had done outstanding work on sunspots. The work of these two men was very encouraging to Morris.

Morris also had the good fortune to have a series of long talks with Stanley Reed who was the sub-editor of the Times in India. Reed told Morris that in Bombay [but not in Calcutta] there was a large class of educated Indians, mostly Parsees, who mixed with Europeans on relatively equal terms; though Indians could not come into a place like the yacht club which was only for Europeans and the occasional American.

Wherever Morris went during this tour of India, he was struck by the lack of scientific journals and reference books. The Royal Asiatics Society's library in Bombay had a fair collection of journals, obtained by exchange, but this was the only library of note west of Calcutta.

While in Poona, Morris had two interesting experiences. First, he noted that the science college, really an engineering college, and the agricultural college were doing fairly good work, which could be enhanced.

Secondly, Morris also had an interesting personal experience while staying in Poona. While at the Club, Morris started to step off the veranda on a lovely, clear star-filled night. He was startled to hear—"Come back"—Morris stopped, turned back and asked the speaker, a stranger,—"Why?"—He replied—"You're just from England aren't you? Snakes!" The stranger then noted that no one walked about in a garden or anywhere outside town at night, without a servant carrying a lantern. Next morning, the stranger showed Morris snake tracks in the dust, which certainly convinced Morris!

Across India to Calcutta

After a visit to see Lord Lamington (the Governor of Bombay) at Maharbaleshwa, Morris returned to Bombay. In St. Xavier's College in Bombay, Morris found able teachers in chemistry and well-equipped laboratories; but the standard of achievement was fixed by university regulations and was at a level that could be easily reached by poor students and be tested by wholesale examinations. The Indian Universities, modeled on the University of London, allowed the poor student, for the most part the product of the poor college, to obtain a degree. As Morris discovered, it was politically unfair to handicap the poor student and the poor college; it was with the best intention that the standard was kept low. After all in India, the possession of a degree gave a man a social status and it was the key to government employment. It was also thought that the number of university graduates gave a reasonable measure of a country's intellectual status and its ability to govern itself.

Morris also discovered a similarity to what he saw in Bristol. He had to look for support for his ideas on education mainly from men in science and medicine and not from men in the "arts".

He also found that the process for ordering books and/or equipment had to pass through three phases: the Head of the College the professor was at; the Director of Public Instruction for the Province and the India Office in London. Often the specifications of the requesting professor were ignored and they were given books and/or materials that someone perhaps 4000 miles away thought best. This often led to items that were totally useless to the professor.

Morris further found that Canning College at Lucknow, Muir College at Allahabad and Queen's College at Benares had well-equipped and well-staffed chemistry and physics laboratories but that the oppressive hand of the dying University of Allahabad lay over all three. The state of education at the Hindu College at Benares was also distressing for another reason. Its Principal Arthur Richardson, who had been Ramsay's demonstrator at Bristol, spoke of the fact that many of his students were poor quality because they were generally married and had become fathers of families before they were sixteen. As a result, their subsequent education markedly suffered.

A.W. Ward, a Professor of Physics, felt that the organization of university education, in general, in India was pitiful and that it would be best to start the Indian Institute of Science with well-equipped and well-staffed departments of chemistry and physics only. Ward felt that there would be few, if any, qualified students in these disciplines available at the start and it would be best for the staff to conduct research on their own. This could serve as an example to Indian students of what they too could achieve.

Calcutta

On December 3rd, 1906 Morris arrived in Calcutta. His first impression was of the great Calcutta "stink". He found a room at the Great Eastern Hotel and saw Risley again who showed him a very large file of papers and a draft of a scheme for the Institute. Morris was surprised since Padshah had told him that nothing had been put on paper for the Institute and that it would be Morris's first duty to advise the government of India as to the constitution and organization of the Institute. To this Risley responded

> "I thought that you would catch Padshah out pretty soon. This is exactly what I should have expected."

Morris here felt that he had Risley's trust and was given some of the papers from the file to study by Risley who asked him to contact him when he had looked over these documents. Risley also told Morris that he might not be facing difficulties of a type that he, and others, had encountered in Europe; in fact there was a set of difficulties unique to India that would have to be surmounted.

With this information in hand, Morris first had to consider the constitution of the Institute. He knew that Ramsay, Masson and Clibborn had advised that the constitution of the Institute should be that of a modern British University, governed by a court and a council. Haldane had basically developed this system. Morris though was not sold on the utility of the court, which in university charters was described as the supreme governing body. The court could, in principle, give orders to the council and the Chancellor that could only be changed by the Privy Council. In practice, the court met only once per year and did useful work in keeping the university in contact with the general public. Clashes with the council just did not occur.

But in Padshah's plan for the Institute (as Morris now learned thanks to the papers he had received from Risley) it was stated specifically that the court, representing government, administration, education, science, technology, etc., through individuals resident in all parts of India should meet once per year. This meeting would not occur in Bangalore but would be in one of the great cities and it would pass the annual budget, make appointments to the staff, make promotions of the staff and act as the executive body of the institution. The council would be a mere management body.

The weak point in Padshah's scheme was this. If the members of it were really able, and thus busy people, they were not likely to spend perhaps four nights on trains to attend a meeting to discuss the workings of the Indian Institute of Science that could be a significant distance from their homes. Either there would be no quorum or a small group could control the proceedings of the meeting. Apparently, this is what Padshah wanted. He could then achieve his ambition of establishing his university of research.

Morris was able, while still in Calcutta, to see Risley again a few days later. He told Morris that he had pointed out the obvious defects in this scheme to the Tata brothers, who felt that it was important to have all of India involved with the management of the Institute. Morris and Risley though clearly favored a council that could meet locally [Bangalore?] for effective management of the Institute. This would prevent the Institute from coming under the control of individuals who were in charge of similar institutions and had "axes of their own to grind". Risley suggested that Morris contact the Tata brothers and discuss the matter with them.

In addition, Morris had the opportunity to meet with two other individuals involved with education in India while he was in Calcutta. One was Archdale Earle (Chief Commissioner of Assam from 1912-1918) who was straightening out matters for the next Director of Public Instruction; Earle was an able man but with no experience or knowledge of education. William Hornell (later to be Vice-Chancellor of the University of Hong Kong from 1924-1937) was acting Assistant Director. Morris saw a great deal of these very able men in 1906 [and 1907] and though Morris gave them advice and assistance about their problems they were not able to assist Morris with his problems.

Before finally leaving Calcutta on December 12th, Morris made contact with the Presidency College and was asked by Alexander Cunningham and C.W. Peake, Professors of Chemistry and Physics, to meet P.C. Ray, who was India's star chemist. Ray's enthusiasm was an asset to India.

Bangalore and the Institute Site

Morris arrived at Madras on the 14th and in Bangalore at about sunrise on the 15th. He obtained a bedroom and sitting room at the West End Hotel; and after breakfast visited the Resident Stuart Fraser [a man possessing outstanding moral and social qualities] who had been a very successful tutor to the Maharaja while he was a child. Since he had won the respect and affection of his pupil there was a good reason for him to be chosen as the British representative to the Kingdom of Mysore [which was possibly the most prosperous and advanced state in all of India].

Fraser opened their conversation by asking Morris what he was going to do and Morris gave him an outline of the agreement for the scheme of the Institute and told him what money was available. Fraser knew about the site; but he was incredulous when Morris told him that the Mysore Durbar had promised a capital grant of Rs.500,000 and an annual grant of Rs.50,000.

The site to be assigned to the Institute was about a mile from the Civil and Military Station. It would come under the jurisdiction of the Resident as Chief Magistrate, though it was not under the Bangalore municipality. Morris had to settle the details and though the arrangements seemed very complicated they all worked out well.

Morris next called on the Dewan, Vishwanath Patankar Madhava Rao who had been a former Dewan of the Travancore State. Morris described him as "a striking looking man with a grave face". Generally, Madhava Rao was a very reserved individual but he was quite warm with Morris and they discussed a variety of subjects ranging from religion to state administration.

The next day, Madhava Rao sent his car for Morris and had him driven to the prospective site for the Institute. Though the location was beautiful, it would have to be completely cleared of jungle and replanted. This would be done later. Close by were the filter beds of the modern water supply system for Bangalore; water was being pumped from a source twenty miles away. This system would supply water to the Institute. Electricity for the Institute would be available after an electric power line about a mile long was connected to the high-tension system from the Swasamudram power station on the Cauvery River.

Having completed a visit of the proposed site, Morris called on McHutchins who was the Chief Engineer for Mysore. He told Morris that, in his opinion, the scheme for the Institute was dead and buried and that he wanted this site for the extension of his filter beds. Morris graciously noted that there would be room for the Institute and the filter beds and that he would discuss all Institute projects with him. McHutchins then wisely noted that it would be best if the Institute had its own architect and contractor; a suggestion that his department build for the Institute would have led to complications.

Before he left Bangalore, the Maharaja received Morris. Morris found his Highness to be a very polished gentleman with a quiet dignity about him; he briefly told him of his tour of India and what he hoped to do. When Morris was leaving, his Highness said he hoped that Morris's visit to Mysore would be a long one; he looked pleased when Morris said that he hoped to spend many years in Mysore.

On December 17th. Morris learned by cable that his father William was gravely ill, and on the 19th that he had died. Fortunately, he was only ill a week before passing away and William had actually seen patients on the day he took ill. In his sadness, Morris said that William was a very good father to him, and that they had always had been good friends, particularly as Morris had a high regard for William as a man.

Back to Bombay

There was nothing more that he could do in Bangalore so Morris left on the night of December 20th for Bombay traveling by the narrow gauge railway which wound its way in and out amongst the hills on the eastern side of the Ghats as far as Poona. At about mid-day on the 22nd, Morris arrived in Bombay. There was no news of the Tata brothers or of Padshah; but Morris had the opportunity to speak with A.J. Bilimoria who was the Head of the Tata Bombay Office. He at once saw the point of Morris's objection to control of the Institute by a body supposedly representing all of India, meeting once a year in one of the capital cities of India and also of giving any degree of control to the existing five universities. Bilimoria told Morris that he was right in attributing the Institute's first draft to Padshah.

Later that same day, Morris had a long discussion with Ratanji Dadabhoy Tata, Bilimoria, other Tata firm members and Tata's lawyer, A.P. Vakil, who, being an Indian university graduate in law, was quite insistent on retaining the representation of the universities. Bilimoria was instrumental in getting the new draft passed. At the end of the meeting a cable was sent by Bilimoria to Padshah saying that the meeting had approved Morris's re-draft, which had also been approved by Risley, and suggesting that the business required to bring the Institute into being should begin. A.P. Vakil gave Morris a letter to this effect; but on the 23rd Bilimoria asked Morris not to proceed until a reply had been received from London. To this Morris agreed. Morris was able to learn that Padshah had written that he hoped that any money that was in excess of the cost put forward in the Masson & Clibborn Report would be used to help establish a department of archaeology.

Morris responded in a long, frank and friendly letter to Padshah explaining his revised constitution. He told him of what he had seen of the Indian universities, colleges and students. The quality of the students was such that the Institute would probably have to give them additional undergraduate training that would call for more staff than had been originally anticipated. There was insufficient money to establish an Institute for Science and Technology even as modestly outlined by Masson & Clibborn and it was much too soon to talk about other subjects such as archaeology.

Morris also had to correspond more with his family because of William's death. Fortunately, his mother Anne been left reasonably well off financially and she decided to stay, for the time being, at 2 Phillmore Gardens while Morris's brother Ernest, also a doctor, picked up what he could of his father's practice. To his youngest brother, Wilfred who had just completed his training as an architect under Sir Aston Webb, Morris was able to help with money. Morris's sister was in Jamaica at this time.

After a painfully dull Christmas that was made somber by William's passing, Morris had a most interesting week renewing and extending contacts made during his first brief visit. Morris had been elected a member of the Bombay yacht club and he lived in the yacht club chambers. Members of the club included representatives of European commercial houses, professional men, members of the press, officials of the government of Bombay, and of the municipality, but relatively few soldiers. No one group dominated it and the club was Morris's pleasant home during the many visits he paid to Bombay. Morris made many friends with all classes of the European community while a club member.

January 1907

Morris returned to Bangalore in January 1907 but nothing official could be done because the Resident and the Dewan were away on state business.

Unofficially, some work was done on adjusting the south boundary of the site to extend the water works; this Morris agreed with. There were also some interesting discussions that Morris had with members of the Mysore Education Service on the future relationship of the Institute and the Mysore colleges. Given that the state colleges were affiliated with the Madras University and that its standards and age of matriculation were to be lowered, it was proposed to form a University of Mysore. Morris's idea to establish a 'college-side' to the Institute was well known; he was here following the precedent of American Universities such as the Massachusetts Institute of Technology (MIT). The question now was—Could the Institute and the new university co-operate? Unfortunately, the latter lagged too much behind the former for the idea to have a chance to actually blossom.

On January 13th, Morris left Bangalore for Calcutta though flooding from seasonal rains forced him to take a circuitous route that put him in Calcutta only on January 16th. This was the height of the cold weather season; Morris could not find a hotel room. Fortunately, his friend Dr. Annandale invited him to join the Museum "chummery", which occupied a house in the museum quadrangle. It was in this bachelor residence that Morris lived during his many visits to Calcutta.

Morris still thought that the Tata brothers, acting as the executors of their father's Will, were only waiting for the amended Institute plan, which after such further discussion as they might desire and consequent agreement on details would be submitted to the Indian government with an order 'vesting' the endowment properties with the Treasurer of Charitable Endowments. Morris had no idea that Dorabji and Ratan still held to the view that they had the right to dictate not only the terms of the constitution but also to insist upon the adoption of an organization that was a complete departure from what their father had agreed upon before his death.

Risley now told Morris that the only possible cause for further delay might lie in the wording of the agreement regarding the transfer of jurisdiction over the site at Bangalore, which was not to be 'ceded', as was the Civil and Military Station. The lawyers might find difficulty here but they did not.

Up until this time, Morris had only visited the large towns and Bangalore. But an invitation from Cyril Bergtheil [later Berkely] and his wife, both former University College London students, took Morris to Behar that was in the district between the Ganges River and the foothills of Nepal. This area was still the home of the Indian natural indigo "industry" which was suffering much from competition with the synthetic German indigo industry and poor organization of its planters to combine, hold stocks and to control prices.

Morris returned to Calcutta on January 31st. He found a letter for him from Bombay saying that the Tatas agreed that he should employ C.F. Stevens & Company, architects from Bombay, to draft plans for laboratories and other Institute facilities. Risley was informed of this by Morris and told Morris to proceed. Morris then wired Stevens to meet him in Bangalore a week from then; Morris then left Calcutta for Bangalore the night of the 31st.

February and March 1907

C.F. Stevens was the son and successor of the architect of the Central Railway Station and other important buildings in Bombay, and was considered to be the leading architect in India at the time. Stevens met Morris on February 2nd and they spent the morning on the site. They agreed on a general layout of the buildings, which was then adopted. They could also quarry gray granite (gneiss) on the site. Stevens put together the cost of labor and materials. They then saw the Dewan, presented their ideas and received his approval. A revised plan of the site was also made which took into account McHutchin's filter beds. All of these documents were then sent to Calcutta.

Travers arrived in Bombay on February 20th and was able to see Dorabji Tata that afternoon. Dorabji told him that Padshah felt Morris had given in to the government too much on the Institute scheme. On the morning of the 21st, they saw Stevens's plans. Though Ratan Tata liked the design of the main building, he was most anxious to cut down the accommodations for the staff; he could not get away from the idea that the staff would be young graduates from the British Universities. Ratan had been informed that the Mysore Durbar was building a bungalow for the Maharajah's secretary at a cost of Rs.45,000 and that Morris's estimate for a bungalow for himself at Rs.25,000 was too high. Morris here told him that the truth was the exact opposite.

The Tatas were anxious that Morris should, at this stage, ask the government for more money for capital expenditures. Morris told them that they should be at least Rs.300,000 short in order to carry out the original scheme. Morris proposed to use the surplus of income, which would accumulate while construction was ongoing as this was the usual practice in the case of university foundations. The government of India and the Mysore Durbar agreed to this and Morris could not possibly imagine that the Tatas would object. But Morris was mystified when they did not like the idea, though they did not express themselves strongly at the time. Later, they took a line on this issue that left Morris stupefied; this change was obviously due to the influence of Padshah.

While in Bombay this time, Morris had a chance to answer a question he himself had about the ability of Indians as craftsmen. In late February or early March, Morris had the chance to be taken on board a 30 ton yawl-rigged yacht built by one of the Wadia family from whose shipyards in Bombay had been launched many of the old Indian tea clippers. The workmanship on this vessel was perfect in every detail and left Morris feeling very good about Indian craftsmanship.

Morris then received an invitation from the Gaekwad of Baroda [whose kingdom was only a few hours rail journey from Bombay] to visit and give a report on the needs of his state with regard to science education. The Baroda College, affiliated with Bombay University, was only a secondary school. The Principal, an able young Englishman, told Morris that the object of virtually all the students was just to get a degree. Any sort of attempt to raise standards would empty the college; the students would just go to another institution.

Morris attended one of the chemistry lectures, which consisted of reading from an inorganic chemistry text and attempting to carry out, quite badly, experiments noted in the book. In the physics department, they had plenty of apparatus though none of it was being used. However, the technical school ran a very practical three-year course in applied chemistry [which had been started by Gaggar]. Students in this school aimed at becoming dyers, soap makers, etc. Morris though realized that he would have to spend additional time in the institution to see exactly why the college was ineffective and the technical school was a success.

On March 20th, Travers returned to Calcutta where he found a wire had just been sent summoning him from Bombay. Risley told him that

"he was in the middle of writing the most important dispatch ever sent from India to the Home Government."

This was the first step in the introduction of the Minto-Morley reforms, embodied in the India Council act of 1909, providing for the addition of elected members to the Imperial and Provisional Councils. Seats would also be provided for Indians on the Viceroy's Executive Council and on the Provisional Executive Councils. This would be the first step towards self-government for India.

But significant as this was, there was no advance directly relevant to Morris's work. He was instructed to wire the Tatas asking them to arrange a meeting at an early date. They stated that their lawyer would come but they could not leave Bombay until after Padshah arrived back in India; this was an ominous sign. The lawyer duly arrived in Calcutta and met with the government's lawyer about the transfer of properties.

April and early May 1907

In late April Morris received a letter from Padshah saying that Morris's amendments to the scheme for managing the Institute:

> " . . . went against the wishes of the late Mr. Tata, and his pledges to the general public. Both Sir Herbert Risley and yourself are more or less unaware of the history of the growth of the scheme. Neither of you know how much public opinion, and even Government, as represented at the Simla Conference, had to be educated to the idea of an institution run at the discretion of the professors. An understanding had to be established with the prevailing opinion by giving as much interest as possible to the culture and wealth of the country on the purely non-academic aspect of the Institute administration, and the voting of the annual budget was therefore entrusted to the Court."

This was nonsense according to Morris. Padshah did not know that, thanks to Charles Martin, Morris had a collection of papers telling the whole story and proving that he was definitely not ignorant of the history of the scheme. He also knew that Ramsay and Masson/ Clibborn had proposed a court and an executive council for the management of the Institute; these individuals certainly knew what the functions of these bodies were in a modern university.

By this time, Morris now truly understood Padshah and what he was trying to do. His objective was to try to have the Institute governed by a body that could be easily prone to political manipulation. Then funds could be diverted to his idea of establishing departments devoted to what he considered to be cultural subjects (such as economics and archaeology to name but two).

Morris realized that arguing with Padshah was pointless and he did not reply to Padshah's letter but instead waited for the Tata's next move. Almost immediately, they submitted their original scheme for the Institute, with additional clauses, placing control more firmly than ever before in their hands. Morris of course received a copy of these papers.

It was already quite clear to Morris that the solution of the general problem of higher education in India consisted in separating the colleges into two classes. One class would provide up to the B.A. or B.Sc. standard and the other class would provide courses for higher degrees. This arrangement would however place the more highly organized colleges in India in much the same position as the recognized schools of the University of London held in relation to their university. This would make it possible for the University of Calcutta to retain the fullest control over the smaller colleges, but to allow the Presidency Colleges, which could admit only B.A.'s and B.Sc.'s, some freedom and control over courses of study.

This system would though need a very powerful man in the government to adopt this as policy and see it through. Morris found out that the word policy was unknown in application to such matters as education in India. Fortunately, Morris was able to find such a man in George Sydenham Clarke who would become the Governor of Bombay in 1908. Clarke was, with Morris's help, able to successfully carry this idea through in the teaching of science in Bombay.

As a part of developing this system, Morris was asked to look at the state of education in the Presidency College. He began his report by noting that the

> "College had no constitution, nor organisation which could ensure the maintenance of a definite policy, and the co-operation of the members of the staff for the advantage of the institution as a whole."

The Principal was solely responsible for internal organization and as there had been four principals during the last four years, it was difficult to see how continuity of policy was possible. The appointment, and pay, of the principal was a matter of seniority in the service, which was extraordinary considering that the office was one of great importance.

Morris also examined the relation of the college to the University of Calcutta, which framed regulations suitable to the needs of a large number of small colleges, but were oppressive when applied to this college. He outlined the possibility of providing greater freedom to such institutions with regard to teaching for higher degrees. This suggestion basically gave the staff of the Presidency College a reasonable degree of academic freedom.

His efforts also found that holding classes in technical areas (e.g., electrotechnics) in association with pure academic disciplines (such as physics) was probably futile. In addition, university examinations also were significantly interfering with the work of the college, so that there was, on the one hand, no long vacation, while on the other hand teaching was confined to only half of the year. This reduced efficiency and damaged morale.

Morris only differed from his friends in the Presidency College on one point. People in the college wanted him to recommend an increase in scholarship grants to senior students. Morris felt as Ramsay did about scholarships; they should be regarded as loans and be paid back. He did not propose to offer scholarships so that students could be induced to come to the Indian Institute of Science. So, Morris, knowing that his views on this subject were not likely to be accepted, said nothing about this issue in his report on the college.

However, just before leaving Calcutta, Morris and Alexander Cunningham had a heated discussion about admission of students to the Indian Institute of Science that might come from the Presidency College. Cunningham became angry with Morris and they parted on bad terms never to see each other again.

Cunningham later was dismissed from the Presidency College due to a letter he had written and was exiled to a college in a town far away, where he died suddenly soon after arriving. Though Morris was criticized for his disagreement with Cunningham, it is likely that Cunningham was discharged for either complaining about his teaching load [he and two others had each to give 54 lectures a week; Cunningham's B.Sc. students only received two hours of lab work per week] or for commenting on how the government was mishandling the participation of students in political disturbances. The government was trying to force the university to take disciplinary action against these students while the university only recognized the individual student as a candidate for examination. Perhaps, Cunningham wrote a letter protesting what the government was trying to do; so much for the academic freedom of a professor!

CHAPTER XI

The Tatas, Politics and the Institute

May, June and July 1907

Morris left Calcutta for Simla on May 19th. Traveling across India in the middle of May was a new and very unpleasant experience for Morris. Driving all day through a gray, dusty haze in temperatures that were still 116 °F in the early evening was unbearable, to say the least. Upon arriving in Simla, Morris was put up at the Grand Hotel where he met Norman S. Rudolf in the smoking room. Rudolf had received a M.Sc. from the University of Liverpool and was a classmate of Francis. He also, as was Morris, born on January 24th, 1872. Morris found that Rudolf had an extraordinary knowledge of India and particularly of its vegetable and forest products as well as the sources and manufacture of many drugs.

The next morning Morris saw Risley who noted how satisfied he was with Morris's recent work. He told Morris that he appreciated a man who was persistent and that he had ordered the complete file of the papers relevant to the Institute to be sent to Morris; this was certainly a sign of his confidence in Morris.

Morris also now learned that the Indian government had never seen a copy of J.N. Tata's will; Risley also said that he would ask the government of Bombay to get a copy of it. Unfortunately, Risley's efforts were not successful and Morris only saw the will in August when he was again back in Bombay.

Risley had, by the way, committed a **serious blunder** and accepted the position of the Tata brothers who felt that they were acting voluntarily in carrying out their father's wishes and claimed to be joint donors with their father on the endowment properties. Before accepting their position, Risley should have had J.N. Tata's will examined by his own legal advisers. This error allowed the Tata brothers to ignore the agreement between their father and the government of India and gave significant justification to their argument that they could decide the direction of and constitution of the Institute.

At this point, Morris offered to resign his position as Director. This would have happened had Risley not assured him that he would have the support of the government in carrying out the scheme that was agreed upon at the time of his appointment.

But poor Risley, already over-burdened with the work of drafting the council's bill, had also been assigned the work of drawing up regulations dealing with the handling of seditious activities by Provincial Governments. Morris here had grave doubt if Risley could find time to attend to affairs of the Institute, particularly because he really knew nothing of modern university education and of the difficulties [that he had created by not having the will examined] which, now had to be faced.

Morris saw Risley again late in the day on June 3rd and he looked exhausted. He told Morris that he had talked to the Tatas in 1904 about the constitution of the Institute and at that time it was clearly understood that the court would be purely advisory with the council being the executive body of the institution. Risley accepted Morris's re-draft and suggested that he give it to Edward Giles (the acting Director General for Education); Morris was happy to do so.

While Morris did not intend to deal with what had become known as 'teaching industries', he wanted a man who understood the practical side of the industries that could be developed in India. Rudolf had not only this ability but also was very familiar with the natural products of India. Morris offered this "position", which still needed to be created formally to Rudolf who accepted. A young chemical engineer would pair well with Rudolf in this "department".

Return to Bombay and Clarification with the Tatas

Morris left Simla on July 29th and drove to Kalka from where he got on a train to Bombay. The train though did not leave Kalka until late at night and Morris slept in his carriage without a mosquito net. He was bitten repeatedly and contracted a dose of malaria, which was to be a recurring problem until he left India in 1914.

Dorabji and Ratan Tata were in Bombay, as was Padshah, when Morris arrived. His arrival seemed to surprise them and they could not, or would not, meet with him at this time. Given this, Morris then went to Poona to stay at Government House with the acting Governor of Bombay Sir John Muir MacKenzie. A topic of discussion involved a report Morris had been asked to write about improving science teaching in the Bombay University. This 12-page report said that making immediate drastic changes would do nothing but a well thought out policy could lead to improvement. The majority of the colleges must actually be regarded as high schools even if they continued to award, for whatever political reasons, B.A. degrees. Serious science teaching must be concentrated in Poona.

It turned out that Morris's report led to a committee being formed by Sir Steyning Edgerly (who was in charge of educational matters) to develop Morris's idea. This ultimately led to the science college at Poona in 1914.

Upon his return to Bombay, Morris soon saw J.N. Tata's will and the whole case that had been put before the Advocate General, Bombay.

Morris soon had a series of meetings with Dorabji Tata and his lawyer that began bitter but ended with civility. In these meetings, Morris, taking the position of the government, wanted to know what was the exact position of the Tata brothers and what was their position regarding the endowment properties. This was done for the benefit of all in order to lessen the possibility of future litigation.

Next, the fact that neither government nor Morris would accept Padshah's constitution was brought up. Dorabji, angry, went into a long story of the scheme but after Morris corrected several of his points, he realized that Morris knew more about the plan than he did. Morris then asked Dorabji who actually had drawn up the scheme and Dorabji responded by stating that Padshah had been sent to Europe to do so and that the scheme had been drawn up after the Simla Conference.

Dorabji also dared to say that the Tata brother scheme had been based on advice given by Ramsay and Masson. That Dorabji was speaking in nonsense was obvious even now to him. He then went off on a tangent and asked why Morris so scorned India's indigenous institutions. Morris astonished him by first showing that the Indian universities were antiquated copies of British institutions and then by reading to him a record from the Institute file in Simla of a conversation between Dorabji and Risley in 1904 where the relations of court and council in the scheme put forth by Morris and his allies had been agreed upon.

After one of these meetings concluded and the lawyer had left, Dorabji asked Morris what was his problem with Padshah. Morris immediately replied that Padshah had lied to Morris when he had asked him if there were any schemes for the management of the Institute that had been drafted. Dorabji replied by stating that Padshah did not want to frighten Morris; and he felt that Morris would fall in with the Tata ideas so he did not bother to show him the agreed upon scheme.

After further meetings between Dorabji, his lawyer and Morris, there was still no agreement though the atmosphere had cleared up and the Tata "side" realized that Morris knew his business. Morris then became ill with fever that delayed matters for ten days. By the 27th, he was again well enough to proceed; he met then with Dorabji without making any progress. Two further days of talks between Morris and the Tata's lawyer followed where Morris presented a compromise that had come to him while bedridden. The lawyer thought the Tatas would agree to this solution, which still needed some revising which Morris did.

This group met again on August 30th with the addition of A.J. Billimoria. The compromise, agreeable to all, was there would be a court and an executive council as seen in British universities. The court could recommend or criticize but could exercise no executive authority nor could it interfere with the executive authority of the council. However, four members of the court should form a standing committee of the court. These individuals were to be selected by the Patron, the government of India, the Mysore Durbar and the Tata brothers. This committee might consider any Institute matter coming to their attention and they could petition the Patron and ask him to look into that concern. Whatever the Patron would find from this enquiry would be binding on all, including the executive council. Unfortunately, this compromise would only work if the Patron acted as protector to the Director of the Institute and if the council members were free of previous agendas and prejudices. With this made public, the Tata brothers promised to hand over the accumulated income [about Rs.550,000] from the properties from the date of the death of their father.

Leave in England

On August 31st, Morris sailed for home on leave and arrived in London on September 16th after a very pleasant voyage. He spent about eight weeks living with his mother at Number 2 Phillimore Gardens where his brother Ernest was in medical practice. While Morris was home, Ernest was married.

On September 22nd, Morris went to Bristol where he stayed with Ferrier for almost a week. While there, Morris was initiated into Freemasonry at the St. Vincent Lodge. Very little had occurred with the Bristol University scheme because the Merchant Venturers were holding out for the control of all of the science teaching in the new university.

Morris went to Liverpool on September 28th for the opening of Donan's new laboratory. Wilhelm Ostwald gave what may very well have been his last address which in Morris's eyes was a pity. Morris then spent an additional four days in Bristol.

In early October Morris had minor surgery on his nose; this would be the first of many nasal procedures. Later that night he had intended to go to a students' dance at the University College. Though he was not feeling well and was extremely shy around unmarried women he did not know well, he was persuaded to go by his mother. There he again met Dorothy Gray who now was nine years older (23) than when he had last seen her after her brother's laboratory accident in 1898. Interestingly, Morris had declined two invitations form the Grays in the last year to again "meet" Dorothy. Morris also had dinner with Dorothy and her sister and saw a play; he also had supper at the Grays. Though he knew that he and Dorothy could not meet again before he left for India, Morris and Dorothy agreed to write letters to each other until they met again [which was almost two years later].

While also in England, Morris found it difficult to get men to go to India unless they were young. For the position of Professor of Electrical Technology, he secured Alfred Hay who had been Lecturer on Electrical Engineering at Cooper's Hill College until that institution closed down. Hay proved to be a good engineer and departmental officer whose accent though (having been raised in Russia) presented minor difficulty in India. Morris later realized that he did not offer enough money to attract the men he wanted (£1000 a year rising to £1250 a year, with 10% for a retiring allowance—the obligation of retirement at age 55 from service in India being a serious obstacle).

Back to India and Further Intrigue

Morris's voyage back to India was pleasant and uneventful and he arrived in Bombay on November 22nd. He called on the Tata'a lawyer that same afternoon responding to a letter from Dorabji who now wanted to go back on part of the agreement, which had just been reached. Morris soon saw Dorabji who was quite friendly though he noted that the income from the endowment funds would not be paid prior to the date of the Institute coming into existence.

This change was due to Padshah who knew that, at this stage, the government would not fight over Rs.450,000 so he persuaded the Tata brothers to hold onto this money so pressure could be put on Morris. Interestingly enough, this money was later offered in the name of Ratan Tata in 1909 to establish a Department of Social Studies and by Dorabji Tata in 1912 to establish a Department of Medical Studies. These two offers, though laudable in their own rite, were packaged with so many conditions that the council had no choice but to reject each offer.

Morris believed the Tata brothers hoped he would fall in with what were obviously Padshah's views. It was also true that neither Tata brother knew much about their father's ideas, since he had never taken them into his confidence. During the meetings of August 1907, Morris had told Dorabji the: detailed facts of his father's ultimate agreement with the government as noted in a letter dated April 8th, 1904; Padshah's about face immediately after J.N. Tata's death and of his conversation with Risley later in 1904.

Amazingly, after Christmas the Tata brothers now offered to pay the guaranteed income from the endowment properties, Rs.125,000 from January 1st, 1908 on the condition that the government of India and the Mysore Durbar paid the grants promised from that date. The Finance and Foreign Departments had to approve this and formal approval took some time.

On January 8th, 1908, Morris went from Calcutta to Bangalore. The Dewan told Morris that "his" grant would be paid when the endowment income from the Bombay properties was received. The two men went over the plan for the site and its buildings only differing in the desire of the Dewan that the Institute have an eastern touch; Morris noted that there was no money for this feature. The Dewan said that he would take an active interest in the Institute, which he most certainly did do over time.

On January 12th, while Morris was in Calcutta, Rudolf and Hay had their names put before the Viceroy (Patron of the Institute) for confirmation of their appointments. Though the Home Member of the Viceroy's Council objected to their hiring stating it was too soon to appoint professors; Morris replied that he could not carry on single-handed and that he did not have the necessary background to design specialized laboratories, such as those needed for electrical engineering.

Provisional Committee of 1908

All involved now agreed that there should be a provisional committee appointed to exercise the full powers of the council so that work might be started on the buildings, etc. After letters were exchanged with Risley, Morris went to Bombay to see the Tata brothers and the architects. He then went to Bangalore, saw the Dewan and dined with the Frasers at their residence on February 11th. This was actually the first opportunity Morris had to speak with Stuart Fraser about the Institute. Morris provided him with an outline of the scheme for the management of the Institute and a plan for the Institute's buildings and facilities. The next day, Morris wrote a long letter to Fraser (who now was very interested in the Institute) giving additional details about it and suggesting a slight amendment to the composition of the Institute's executive committee. Instead of having the Director of the Institute be the council chairman, Morris suggested that the Resident be its chairman.

The provisional committee of 1908 consisted of: Stuart Fraser, Resident in Mysore; V.P. Madhava Rao, Dewan of Mysore; Morris Travers, Director of the Institute; Alfred Hay, Professor from the Institute; Norman Rudolf, Professor from the Institute; Burjorji Padshah, nominated by Dorabji Tata and H.J. Bhabha, nominated by Ratan Tata.

This committee first met in Bangalore on March 23rd, 1908 to consider a report that Morris had been recently been working on. All members had a printed copy of this report as well as the Tata brothers, the government of India, the Mysore Durbar, the leading Indian newspapers and the London Times. The report seemed to be well received and the London Times devoted a long article to it, including a good summary of the history of J.N. Tata's scheme and a reasonably accurate statement of Morris's proposals. A quotation from it stated:

> "The chief work of the Institute will be the establishment of departments of pure and applied science, and students who have passed through the Indian Universities will be trained so that they may apply science to the Indian arts and industries. It will be in no sense a trade school. Though there will be no undergraduate side at present, it is expected that this may ultimately become necessary, as has been the case with some projected post-graduate institutions in America. Even as it is, most of the Indian students entering the Institute will have first to go through a course of practical instruction before commencing research. Private workers requiring accommodation for the purpose of investigating new products or processes, or activated by a desire to carry out scientific work, will be received. Six departments are to be established, each with a professor

and assistant professor. The Director will occupy the Chair of General Chemistry, and a Professor of Applied Chemistry has already been selected. In view of the importance of vegetable products, there will also be a Chair of Organic Chemistry. The nearness of the great Cauvery power works, from which a supply of electricity at high tension will be obtained, has led to a decision to open a Department of Electrical Technology. There will be a Chair of Applied Bacteriology, a sixth department has not yet been finally decided upon. A large sum has been allotted to the creation of a library. Probably 60 students will be admitted to the Institute in the course of the next two or three years . . ."

Morris mused that if he had written this, he would have stressed the importance of establishing nuclear departments in chemistry, mechanical and physical science and biology.

Only the Resident, the Dewan, Hay and Morris were able to attend this first meeting of the provisional committee though. No Tata representatives were present though Dorabji acknowledged receiving the report. At the meeting, Fraser and the Dewan asked Morris several questions and they were generally pleased with the content of the report. The report was then placed on record. The committee then looked through the construction plans for the Institute and approved them and asked Morris to have the architects draw up estimates and advertise for bids for the construction of the foundations and plinths of certain buildings. Following this meeting, Morris and the Dewan agreed on a simple classical style for the Institute buildings due to cost [though the Dewan favored an oriental style of building].

Building the Institute—The Beginning

The main building, housing the library and administrative offices, faced due north onto an open space. The first laboratories to be built were to the east and west of this building and were arranged so that additions could be later made to them, if needed. Space for other laboratories was reserved to the south of the main building.

Bungalows for the Director and staff surrounded the open space. Blocks of students' quarters were to the west. These student accommodations provided a separate room for each student and a mess room and kitchen for each of 12 students. Kitchens and mess rooms were as far apart as possible. 24 students shared dormitory style bathrooms. Morris was also sensitive to the fact that he would be drawing students from all over India and that prejudices as to feeding for a variety of races and castes had to be considered.

The subsoil was gneiss, a gray granite-like material that could be used to help build in a style of mainly coursed rubble masonry. Trial pits sunk about the site revealed that immediately below where the tower of the main building would be erected there was a small pocket of china clay, formed by the action of water on the gneiss. The tower was then built on a platform of light steel beams on concrete.

Rudolf, before leaving in February 1908 for England, had advertised for a man to act as engineer in charge of building works. He received many applications, mainly from former government officials, but he selected a man named Miller who was a young Englishman who had been trained in his father's building business and had spent significant time in India. Miller was an excellent draftsman and had experience in keeping building accounts and records but not with the public works system.

Difficult as it was to predict the cost of building an institution from scratch, Morris set down a capital cost for buildings and equipment at Rs.1,300,000 where Rs.750,000 had been allocated to the Institute from the government of India and the Mysore Durbar. Morris proposed to make up the deficit of Rs. 550,000 from savings on income during the period of development, a procedure that was common in universities in Europe and America. There was also financial assistance coming from the governments of Bombay and Madras so Morris was not concerned. In addition, there was no opposition to Morris's handling of finances during the March 23rd meeting.

Morris's Office, Staff and the Grant

Morris now was able to open an office in Bangalore. His chief clerk was the young, well-educated Sundaram Iyer. An accountant, Gundu Rao, and an assistant for him, Bashirodien, were hired soon afterward. The only common language that these men spoke, as true with many other Indians, was English; they spoke and wrote it as well as Morris. Morris's staff also included two orderlies, Ramaswamy and Swilingum (both old soldiers from the Madras Pioneers) and a number of coolies and sweepers.

Rudolf suggested that all the lower grade workers be paid at a fixed monthly rate plus a bonus for good service and conduct. This bonus though led to many petty complaints made by one worker against another. Morris, who did not wish to waste his time dealing with minor complaints, summarily pronounced to all that the bonus was cancelled for a month. The punishment was actually remitted on payday, which Morris typically handled, and the petty complaints ceased. Morris also held a monthly meeting where he would hear his employee's thoughts, concerns and suggestions. This practice led to his employees becoming very attached to him and exhibiting great loyalty to he and his family before they left India for good in 1914.

On the morning of March 31st, only a few days after his office had been opened, Morris was informed by Iyer that

> "Sir, I hear that there is a large sum of money for you in the Resident's Treasury, and that if you don't take it away today, you will not have it."

Immediately, Morris went to the treasury and was told that they had Rs.87,000 from the government of India for him.

The money had to be taken in the morning since the afternoon was a holiday and the grant would lapse as all grants do at the end of a financial year. Furthermore, the money had to be taken in silver rupees so Morris drove to the bank, got the help of a couple of clerks and took away the boxes of rupees in bullock carts.

Morris banked all the money though in a special account and even made good the value of five rupees that were bad; he also decided to talk to the Finance Department about this amount when he next visited Simla or Calcutta. Then, he fortunately saw Baker who told him that he knew the Institute was short on capital funds so he had kindly made the full amount of the grant available early!

April to July

On April 10th, Morris traveled to Bombay where he lunched at Government House and talked with the Governor about elementary science education. While also in Bombay, Morris saw Dorabji who though had no comment on Morris's report. Time with the architects was also necessary to discuss estimates for Institute buildings; Morris planned to bring this information before the provisional committee at their next meeting.

Right before going to Ootacamund for the provisional committee meeting though, Morris had dinner with Burns and ate something that gave him food poisoning and made him ill for several days afterward. He was still sick for the provisional committee meeting on April 23rd.

At this meeting, the only people present were Fraser, the Dewan, Hay and Morris. The Dewan was anxious that the building contract be given to an Indian contractor, but a bid by Skipp was the best received and was accepted. Given that the foundations were to be finished by September, the committee agreed to purchase an engine and mortar mill and rent it to the contractors so that the work could be hastened.

Morris settled down in Bangalore after this meeting. He rented a bungalow that was built of unburned bricks with burned bricks around door and window openings and covered with lime plaster and painted white. It was flat-topped with a pillared veranda. Though it was quite comfortable and looked palatial when seen from the garden, Morris always feared that the structure would somehow collapse.

He also bought a buggy, which held two and a lively horse that was too much for him to ride, named Marvel to "drive" the buggy. The stable attendant had a donkey and Morris had another horse named Peter Pan that he used to ride out each morning to the construction site.

Morris went through the routine of paying calls between 12 and 2 P.M. but his contacts were almost always with men at the club and in the military messes. Though he played tennis badly, he did have the hard tennis court at West Bank put in order and gave tennis parties. The social function that he liked best though were his Sunday evening bachelor parties. This was no surprise in any way given Morris's extreme shyness with young, unmarried women. He also was not above creating social mayhem. In July, Morris gave a dinner to all the girls and boys (18) in the 'station' and was later told that all of Bangalore wondered why. Morris noted that Bangalore was " . . . a great place for old cats; and one must give them something to speculate about."

At about this same time, Morris finally heard from the Tatas who, in replying to a letter Morris had written on December 9th, 1907, said that they proposed to pay the guaranteed income from the endowment properties from January 1st, 1908 " . . . in order to give the Institute a straight-forward start . . ."; however they now stopped payment after May 1908 claiming a refund of Rs.37,000 for payment of Morris's salary and expenses during the previous year.

Morris, in a letter to Ramsay on July 18th, said that at a recent meeting of the provisional committee, Padshah made a ridiculous comment

> " . . . the expenditure of surplus income on building etc., was contrary to
> the intentions of the late Mr. J.N. Tata."

When asked why the Tata brothers had not brought this up before, he had the nerve to say that they had not thought it necessary to 'interfere' at an earlier stage. When Morris asked Padshah what the surplus income should have been spent on, Padshah said, "that it might be spent on subjects that did not require buildings."

September to November

In September, Morris attended an Education Conference at Ootacamund organized by the Madras Government. Though he had intended to keep clear of Madras education politics, Morris had to obtain a grant from Madras and he had to accept an invitation, which came personally from the Governor. The Governor treated Morris as the most important visitor, placing him on his left at a formal dinner. This angered the education officials and they would later show their wrath.

On October 6th, there was another meeting of the provisional committee in Bangalore that Padshah missed. He had written to the committee in advance of their meeting formally complaining that he had not been able to obtain information regarding buildings, etc., from the Director and asked to be placed in direct communication with the architects. Morris was told to reply, refusing this request saying that the provisional committee was unaware of occasions where its members had been unable to receive information that they had asked for.

Also in early October, upon returning from a ride with Rudolf, Morris found a note from Fraser asking that Morris call him *immediately*. Fraser had received a letter from the government marked very confidential and he read parts of it to Morris. Evidently, Dorabji had been to Simla and had created a giant stir. In the first place, he objected to spending more than Rs.70,000 on buildings and equipment. He told the government that Morris had (1) been trying to make him pay up the arrears of income from the Tata estates, (2) refused to let Padshah see the plans for the buildings and (3) not made use of the assistance they had been so willing to render him. The government was extremely annoyed or perhaps only pretended to be; they probably only wanted to end discussion with the Tata brothers, till the properties in Bombay had been handed over.

The only statement that had any validity was the first. When Dorabji had sent in a claim against the Institute for Rs.37,000 for expenses up to January 1908, the provisional committee had correctly referred this matter to their lawyers who had been sent all relevant papers including J.N. Tata's will and the opinion of the Advocate General, Bombay. The lawyers here advised if the income from the endowment properties did not belong to the Institute from the date of J.N. Tata's death (May 19th, 1904) then it did so from the date at which the Institute came into being, which was, at the latest, the date of Morris's arrival in India (November 15th, 1906).

Government now told the story that J.N. Tata had given the money to found a university, so the validity of the will did not arise. This matter was considered closed by the government. It was clear from the letter of April 8th, 1904 that J.N. Tata had given up the idea of a university. Unfortunately, Morris was, as usually occurred with him, to pay the price of governmental bungling. As to the second and third points, Padshah had refused to meet with Morris in Bombay and the Tata brothers and Padshah offered no assistance; in fact they only criticized everything.

On November 12th the provisional committee again met with all members present. The main agenda item for this meeting was estimates for superstructure. Stevens, the architect, had been coming to the meeting but Morris, by wire, told him not to come. The estimates were then withdrawn.

Fraser then read excerpts of the letter he had received from the government. It was noted at this point by Fraser that the government did not wish to express any opinion on the Rs.750,000 plan or on a more extensive one but it was their wish that decision on all such matters should be made by the permanent governing body [executive council] of the institution which would hopefully be appointed in a month or two.

Padshah had apparently come to Bangalore thinking that his views had been accepted. He could not see that the cost for three chairs was not just half the cost of providing for six chairs. When he was challenged by Morris to propose an alternative he initially would not answer Morris but soon noted that there might be several alternative schemes.

The Dewan introduced a resolution, which was approved, that adopted the instruction of government regarding expenditure. Padshah then moved that the Director prepare a scheme for laboratories, equipment and residential quarters for the three professors already appointed. Morris then pointed out that he had come to India to organize an Institute of Science and to do what Padshah advocated would be a waste of his time. Padshah and Bhabha voted for Padshah's idea while Morris, Fraser, the Dewan, Rudolf and Hay voted against it.

In the afternoon, Morris drove Padshah out to the site though Padshah showed no interest in what was being done. The next morning Morris and Fraser rode out to the site. Fraser was anxious to have another meeting of the provisional committee for additional discussion though he eventually realized that it might be best to avoid further quarrels with the Tatas, which would obviously come about in a meeting. That same afternoon, Rudolf drove the Dewan out to the site. The Dewan said to Rudolf that it was impossible to deal with Padshah as long as he would not tell the committee what he was thinking.

Morris had frequent contact with Fraser and the Dewan over the next few weeks. While all hoped that the executive council would come about in the next month or so, no one really expected it to happen so soon. No further requests to hold another meeting of the provisional committee were made, even by Padsdhah.

End of 1908

The work on the foundations and plinths of the main building was continued and it was completed in January 1909. Then, in order to allow Skipp to keep his staff together, and to give employment to Miller and the supervisory staff of the Institute and to speed up the work generally, a decision was made by the finance committee (composed of the Dewan, Hay and Morris) to proceed slowly with the superstructure of the laboratories. To this "change" no one objected.

Morris also paid several visits to Bombay to discuss building concerns with the architects. He also saw a great deal of the Governor who was attempting to reorganize science teaching in the Bombay University. Morris offered his assistance on a personal but not professional level. Morris did not want to take an active part in provincial university affairs and risk angering members of the education service.

The government of Bombay promised Morris Rs.100,000 towards the building fund and later the Madras government gave Morris a grant of Rs.150,000 though Morris was careful to not concern himself with Madras University and technical matters. He also had a great deal of correspondence on educational and technical matters with other provincial governments, but none of them gave him anything.

At Christmas time Morris went to Calcutta to attend the meeting of the Board of Scientific Advice for India. An official gave a brief account of some of the activities of the government and the whole meeting took all of an hour. Morris dined on Christmas day with his friend Annandale after he had made a round of the departments that dealt with affairs of the Institute.

CHAPTER XII

Marriage, Home Life and Institute Affairs

India—First half of 1909

The foundations and plinths of the main building had been completed in January and work on the superstructures began though it was done without a contract. Morris continued Institute business by purchasing books, apparatus and machinery though the delivery of these items would likely take six months or longer.

The endowment properties were finally transferred in May after a delay of six months. But even after the transfer of these properties, further delay occurred, as the government decided to conduct the correspondence related to electing members of the court and the council. Upon reflection, Morris now believed that it would probably take quite a while to fill out the council and the court.

Morris also attended the Maharajah of Mysore's birthday celebrations in June and spent a very pleasant week in camp outside Mysore city.

Leave and Marriage

On July 1st, Morris left India on three months leave. He arrived in London on Sunday, July 18th and went to stay with his mother and his youngest brother Wilfred at 43 Warwick Gardens. The next day—Monday, Morris found Robert Whytlaw Gray [now on Ramsay's staff] who told Morris that his father, mother and sister had come home from their world tour and had a furnished flat in York House in Kensington.

Morris then called on Dorothy, who had exceptional social skills, and they talked in person for the first time since 1907. Almost immediately afterward, Morris asked her father for Dorothy's hand in marriage. Mr. Gray approved Morris's request and the two men agreed on a dowry that would be paid to Morris after Mr. Gray's death.

On the 20th, Morris returned and he and Dorothy became informally engaged; they would become formally engaged a week later during a walk through Burnham

Beeches. Engagements in 1909 were nothing like that of today [2011] and were beset with formalities, a chaperone accompanying the couple even to the theater. There were of course many visits to halls, dressmakers and relatives.

On Saturday, September 11th they were married in a full ceremony at St. Mary Abbot's church in Kensington [Morris aged 37 and Dorothy 25]. They left afterward for their honeymoon in their new Vauxhall car, which they had only received the day before. They had a nine-day honeymoon spent mostly in the New Forest and Dorsetshire, returned to London on the 20th and left for India on September 30th.

Back in India and the Council

Morris and Dorothy arrived in Bombay on Friday, October 15th and stayed there over the weekend. During that weekend Morris met briefly with the Institute's architects and showed Dorothy some of the wonders of India. The happy couple arrived in Bangalore on the 20th.

On July 30th, the names of the members of the court and the nominated members of the council were published. Rudolf, acting as Director while Morris was on leave, had wired Morris and told him this; Morris replied by suggesting that the council, similar to the 1908 provisional committee [with the exception of a new Dewan—an Ananda Rao] should meet at once to conduct business. On September 6th this body met and elected Fraser as chairman and confirmed Rudolf's status as Officiating Director. Padshah had sent a letter questioning the validity of the council to conduct business, as it was incomplete and the court had not yet been composed with its three representatives; it was ordered that no action be taken on this letter. Also, the council reappointed the finance committee with powers to

"receive monies and to invest unapplied income, and to exercise such financial powers as the Council might delegate it."

While Morris was away, Stevens, the Institute's architect, had visited Bangalore and had a meeting with Fraser and Rudolf. They had a general discussion about Morris's plans for the Institute and came to the realization that they were building a small town, complete with services, roads, drainage, water and electricity, etc. It was not possible to foresee all details and exact cost estimation was impossible. Morris reported this at the council meeting of November 4th and moved, with the Dewan seconding, that the council now had the power to conduct business.

The first agenda item before the council was to legitimize all that had been done by the provisional committee of 1908 and by the Director when that body was not yet in action. Both of the Tata representatives were present but they did not offer any explanations for the actions taken by the Tata brothers. The council agreed to legitimize the actions of the Director and the provisional committee of 1908 by a unanimous vote.

A second item for business for the council was to consider a report by the Director, 70 pages in length and complete with illustrations, that dealt with the general organization, subjects to be provided for, and the estimated cost of building, works, equipment, etc. Morris's report noted that the projected cost of a five-department Institute was Rs.1,900,000. Out of this amount, Rs.1,573,000 would need to be available immediately for the cost of buildings, works, equipment, etc., while another 327,000 would not be needed immediately. This would lead to the completion and opening of the Institute by July 1st, 1911. Also included was an amount that would allow 14% for excess over estimated cost.

Morris also wanted to have, in addition to the finance committee, a buildings committee. Unfortunately, the council did not agree to this request. The Tata representatives and the two members from Mysore were strongly opposed to this idea, so Morris could not press the matter further. This omission was to become a critical at a later time.

There was also a statement made by Morris, which was recorded in the minutes of the meeting that would become important at a later time. It was

" . . . that when the Auditors' Report had been received I had found, on examining it, that there was a serious error in it relating to the statement of account for the year 1909-1910."

The fact was that the senior Institute staff kept their accounts on a simplified double-entry system that was commonly used in commerce, with first entries being made in a 'journal'. Morris's accountant understood this and so did Morris and Miller. This system was labor intensive but was practically foolproof. In government accounts all money transactions were supposed to involve cash payments. The official auditor had apparently been attempting to base his audit on the cash book. Though Morris tried to educate him on how they did their bookkeeping, he did not understand the principles of double entry and Morris's efforts were futile.

Ultimately, the Institute was ordered by government to adopt their system of accounts which involved making large numbers of pencil notes in the books that were later erased. Interestingly enough, other members of the government criticized this ridiculous system!

On the following day, November 5th, more business was conducted at another council meeting. Bids for the construction of buildings were considered and the bid by Skipp was accepted. Morris was authorized to sign a contract with Skipp, which he did on November 29th. At once, Morris issued orders to give priority to work on the laboratories, on one 'block' of the students' quarters and on the drainage, water, and electricity supply scheme, so as to enable Morris to admit students and get to work as soon as possible.

Beginning of life with Dorothy

To say that Dorothy was shocked in her first inspection of the Travers bungalow in Bangalore was an understatement. There were no curtains on the windows, the walls were a very rough plaster and painted white. The floors were of rammed dirt covered with palm-frond matting. Dorothy was also shocked at her first experience with Indian toilets.

But Morris had done some good things with their home. He had bought two beautiful Amritzar rugs, which covered the floor of the sitting room, and he had Indian rugs in the dining room and bedrooms. Morris had designed the dining table, chairs and dresser for the dining room and had this furniture made in Bangalore. The house was also well provided with china (Danish porcelain) and silver and their straw-stemmed wine glasses were handled with great care. Morris's wedding present to Dorothy was a Bechstein six-foot boudoir grand piano, cased in brass bound solid mahogany. It soon arrived and relived Dorothy entirely of the solitude which British wives suffered from in India; with most of them playing cards.

The Travers were quite popular in the Bangalore social scene now with Dorothy present and her superior social skills had them dining out often and hosting many dinner parties.

Dorothy also made a good start with the servants, dismissing a ferocious looking Muslim butler who was disrespectful, and letting the cook, who went to the bazaar each morning to buy perishables, know that his accounts would be checked and that his free run of the stores was over. The household also had a butler, and a house servant that looked after their rooms but had no governess. The Travers also had an attendant for each of their ponies, a gardener and two coolies and the inevitable sweeper woman to attend to sanitation concerns. When their Vauxhall car arrived from England, they also had a motor boy who washed the car and polished the brass work every day. These servants all liked and respected Dorothy and the Travers soon collected a retinue that remained with them till they left India in 1914.

Christmas 1909 was on a Sunday and Morris and Dorothy had 16 people to dinner. On the 26th, Morris and Dorothy left Bangalore for Madras where they lunched with the Governor the next day. In addition, Morris attended the meeting of the Board of Scientific Advice for India in Calcutta and advised Sir John Hewitt, the Lieutenant Governor for the United Provinces, on educational and industrial problems.

In Calcutta they stayed with Dr. Annandale in the Museum bungalow, which Dorothy found to be lacking, to say the least. Their bedroom was furnished with the usual bed, dressing table, and a couple of chairs, but the floor was matted and there were no rugs. There were no sheets on the bed and no toilet cover. Of course there were no curtains. From Calcutta, they went to Benares and toured the usual round of the temples and the ghats, which made Dorothy ill at the sight.

After a brief stay in Allahabad with the Governor, Morris and Dorothy went to Cawnpore where they were both struck by the filth of their hotel and the melancholy nature of the place. While there Morris spent some time going over the various mills, leather, wool and oil industries to get facts for a report he was preparing for the Secretary of State for India. Morris did not know that this report would be sent back to the Government of Madras; as a result Morris became deeply involved in Madras politics.

Following this stop, Morris and Dorothy spent the better part of a week in Delhi and Agra. Leaving Dorothy in the very comfortable quarters in a Delhi hotel, Morris went to Rurki for two days to stay with Colonel Clibborn [one half of the Masson and Clibborn report] who was the Principal of the Engineering College there.

Dr. Hankin, a government pathologist, made their visit to Agra particularly noteworthy by giving them a personal tour over Futtehpore Sikiri, the city that Akhbar built but was sparsely populated at present. This visit also brought out the fact that architecture in permanent masonry was confined to palaces, temples and tombs while domestic buildings were built of unburned brick bonded with mud plastered over.

From Agra they returned to Bangalore through Bombay having been away for approximately three weeks.

Institute—First Half of 1910

The council had been bombarded with letters from Padshah containing legal opinions supporting his contention that the council could not transact business until the court had its three representatives appointed even though the council had already said no to him on this matter. However, in February 1910 the newly knighted Sir Dorabji Tata, went to Simla and made a personal complaint to the government. As a result of this, Fraser received a letter from the government that he gave the substance of to a meeting of the council on March 9th, 1910. Tata's complaints were:

i. The Council had been attempting to claim the income from the endowment properties from the death of Mr. J.N. Tata, instead of from January 1st, 1908.

ii. The incomplete Council was not qualified to act and

iii. The building plans were extravagant; particularly that stone was being used instead of brick.

At their last meeting, the provisional committee of 1908 had already accepted the decision of government that they were not to raise the question as to when the money from the Tata estates became payable. Padshah knew this, and since he represented Dorabji Tata on the council, it is likely that Dorabji knew of it too. Also, where possible in the buildings brick and not stone was being used in order to *save* money.

Morris and Fraser agreed that government was just attempting to placate the Tatas and that it was pointless to argue further about this matter. The council actually hoped to deal with this difficulty by passing a resolution authorizing the finance committee to act on its own behalf. However, when Morris asked the Dewan to attend a meeting of the finance committee, he said, quite rightly, that if the council had no power to act, neither had the finance committee.

At this point, building was proceeding rapidly at a cost of Rs.30,000 per month and wages and salaries were an additional Rs.7,000 per month.

Morris also had no idea how long it would take to complete the council; it was March 1910 which was 15 months after the provisional committee had been abolished yet the council still did not exist. So the only legal action that Morris could do was to inform Skipp and members of Morris's own staff that the Institute would stop cash payments indefinitely. Morris would not appeal to the Tatas because that would only involve him in bargaining with Padshah.

Morris, being a man of honor, decided to try and save the situation. He wrote an official letter to government saying that he saw a way of carrying on unless he was prohibited from doing so; though he said nothing to Fraser or the Dewan about this letter. Morris then called on his friend, Sydenham Clarke who was Manager of the Bangalore Branch of the Bank of Madras, and explained the whole situation asking for a loan in *his own name* to the amount of the deposits held by the bank on the finance committee's account. To Morris's astonishment, Clarke informed Morris a day or two later that the Bank had granted his request. Morris now had enough money, Rs.200,000, to carry on till the completed council met on July 13th, 1910.

Affairs of the Council

The new members of the council were the Director of Public Instruction, Madras (D.P.I.), the Director of Industries, Madras (D.I.) and the Director General of Observatories (D.G.O.). Given that the Dewan was new and that Stuart Fraser had gone home on leave and when he returned he would be Resident in Hyderabad, chaos now reigned on the council. The council now was composed of four different groups rather than one homogeneous body and it was:

i.	The Director	1
ii.	The professorial staff	3
iii.	The Tata representatives	2
iv.	Government officials	5

Practically half of the members were in the last category, and a majority was in the last two categories.

Upon Padshah's resignation from the council, the Tata brothers appointed a lawyer named J.D. Ghandy to succeed him.

The D.P.I. had been trained as a zoologist, was a fellow of the Royal Society, but had devoted himself in India to the study of primary education. He had been a member of the Royal Commission on the Indian universities, which Lord Curzon appointed in 1902. This was no recommendation though for a seat on the council in Morris's opinion.

The D.G.O. was born in 1868 and was Head of the Mathematics Tripos at Cambridge in 1889 and became a fellow of Trinity College and a F.R.S in 1904. His headquarters was at Simla and for him to visit Bangalore meant eight nights spent on the train. Occasionally, he made a tour of India and on one such occasion he did visit Bangalore. Morris regarded him as the most eminent scientific man in India; but he unfortunately could not give the council any useful service.

The D.I. was a member of the Indian Education Service and had held the post of professor of engineering in Madras. For several years he had been Director of Industries in Madras. His duties were two-fold; on one side it involved the study of the development of new industries such as the utilization of jungle timbers, unsuitable for building, for the production of charcoal by distillation. His department also aimed at developing industries such as the use of aluminum for hollow ware manufacture and the chrome tanning of leather.

In 1900 Ramsay had been attracted by this idea. In recent years it had been proposed to establish a similar department in the united provinces. However, the Secretary of State for India intervened, prohibited the venture and closed the Department of Industries in Madras. He ruled that the work was outside the sphere of government activities. At this stage the D.I. became Director of Industries in Mysore State and was at that moment elected to the council. Realizing that, as he differed diametrically from the Institute's ideas on industrial development, he told Morris that he proposed to resign from the council; but he did not do so.

The idea of a professor being an independent thinker who was head of his department was unthinkable to the members of the council; they wanted Morris to report in detail on how each of these professors was spending his time. The council, with the D.I. as the lead, also wanted committees of the council to control the activities of the professors. Morris took the opposite side on this critical issue of academic freedom. He believed that after a professor was appointed, he must be regarded as an expert responsible for the conduct of his own department. The work of the departments was to be coordinated by the senate. The council's work was general policy and administration, subject to review by the court through its standing committee, the members of which would almost certainly be completely ignorant of any matter referred to them.

Even though the council met four times a year, six hours devoted to business each time, this was not enough time for effective management of the Institute, even if all members were at each meeting or kept in touch with the Institute, which they did not do. The new Resident (who now had little responsibility) and new Dewan never did take much interest in Institute affairs. The council also was seriously lacking in that it had no architect, civil engineer or businessman on it.

Morris and Dorothy's family—the beginning

Morris and Dorothy, both suffering through the very hot weather of 1910, were both quite happy but concerned. Dorothy was expecting their first child at the end of July. All went well and their daughter, Dorothy Mary, was born on August 1st, 1910. Though there were no gynecologists in India, a Colonel Hudson of the India Medical Service and a well-trained nurse were quite skilled in handling the delivery and immediate follow-up care of mother and daughter.

However, there was trouble when this nurse left at the end of August. Acting on a recommendation from a friend, Morris and Dorothy hired an Englishwoman to care for Dorothy Mary. However, when her mother found out that the child was ill, Mrs. Travers checked the weekly weighings and found that they were being faked. Further investigation revealed that their friend had fired this woman for gross misconduct. Morris and Dorothy then hired a Eurasian nurse who remained with them until Morris had leave in 1912; she only lasted two weeks in England before returning home to India. When coming back to India after Morris's leave was done in 1912, they brought with them a Scottish nanny who remained with the family for many years.

Padshah and the Completed Council

While Morris was at Ootacamund in April 1910, he received a letter and a large number of papers (consisting of 5,000 words of generalities with nothing noted about staffing, organization, etc.) from Padshah. This letter noted that Ratan Tata had been told that a school of social and economic studies could be started in India at the cost of Rs.55,000 per year. Ratan was prepared to contribute a sum of Rs.25,000 per year for ten years on the condition that the remaining amount be made up from the income of the Institute, and that the Institute should provide the necessary capital for buildings, library, etc.

Until now, Padshah had opposed every idea Morris put forward but had never committed himself to a concrete proposition. This new suggestion was in line with creating a school of pure and applied science, but it was clear to Morris [and others] that only Rs.25,000 a year would not allow this scheme to be realized. Morris felt that

Padshah realized this also. Even though Morris did not support this idea, he sent the copies of Padshah's letter and the papers that discussed it to members of the council without comment; Morris did note to the council that they would discuss this at their next meeting.

None of the council supported this idea; indeed several members suggested that it would bring an unwanted political element to the Institute. But a suggestion by Morris was passed and it read:

> "That the Council express their great appreciation of Mr. R.J. Tata's munificent offer, and their admiration of the broad spirit in which the idea of a School of Social Studies was conceived. They consider that the project would be materially advanced by inviting some distinguished sociologist to report on the possibility of successful working of such a scheme, and would be glad if Mr. Padshah would enquire as to whether Mr. Tata would be willing to bear this preliminary expense."

As Morris had expected, Padshah did not reply to Morris's letter. Morris firmly believed that Padshah thought of himself as the likely head of this new department.

When the completed council met for the *first* time in July 1910, Morris had been in India for 3 years and 9 months. It had been 2 years and 7 months since the provisional committee of 1908 had first met. There had been actually three different governing bodies, each of which had been dismissed without adequate reason and in each case causing of significant confusion.

This first meeting of the "completed" council was held at the Residency in Bangalore on July 13th, 1910 beginning at 10 A.M. All the members of the incomplete council agreed to attend, and of the new members only the D.P.I. would attend though the D.I. would be represented by one of his staff.

Morris had informed all that he would be on the site that morning between 8 and 9 A.M; only Fraser and the Dewan appeared. Fraser here warned Morris that the D.P.I. intended to attack Morris's scheme, particularly about the buildings and that Morris had never let him see the buildings or even the opportunity to talk about them. Morris, very angry, responded by saying that the D.P.I. had three chances to do this and had not taken advantage of any of them. Nonetheless, Fraser noted that he understood but warned Morris to be on guard during the meeting.

When the topic of buildings came up at the council meeting, the man representing the D.I. noted he had arrived late at the site in the morning and had gone over the work alone with Miller, who was in charge of it. This man, an engineer, understood and highly appreciated what was being done. He also spoke of "the efficiency and economy with which the work was being carried out". He was also impressed that the walls were of the thickness needed to carry the loads upon them, which was a factor commonly ignored in India.

Years later, Morris learned that this 'late arrival' was not an accident and was actually planned. This was done so that the young engineer could make a thorough inspection of the buildings with Miller alone. He had been instructed by the D.P.I. to do this and to give him his report personally. The D.P.I. stated this was done to curb the extravagance of the Director when in reality the D.P.I. had likely been in correspondence with Padshah.

When the council meeting began, Morris made a brief reference to what had been done during the interregnum. There was no discussion nor any questions on what Morris said. The council passed a resolution confirming all that had been done by previous governing bodies and the Director.

Next, the council confirmed its acceptance of the 70-page report [prepared by Morris] that had been before the council last November and the contracts that had been approved and signed by Morris. Again, there was no discussion.

Since Fraser was leaving Bangalore, the council thanked him for having been its chair. He replied by saying that he was happy to see the council now finally established. It was now only a matter of time till the Institute would open.

Institute 'Construction' Work

When the south wing of the electrical laboratory was finished, bookcases, built to Morris's own designs, were installed in it. The idea that books required careful storage was new to India; in fact one day when Morris was showing Bhabha around the library thinking that he might be interested, he was distressed to hear Bhabha ask why the Institute needed all those books. This also came up at the next council meeting where Bhabha entered a protest against the purchase of expensive bookcases since "any kind of almirah [cheap case] was good enough to keep books in". In fact, Morris was to be told in 1950 that in establishing this library, he had conferred an "inestimable boon" on southern India; bookcases were still being made according to Morris's design.

At the time Morris accepted the directorship, he understood that the construction and equipment of the buildings of the Institute would be carried out by the Mysore State Public Works Department. Morris had found out that this department could not help at all; and on the advice of the Tata office the work was placed in the hands of a Bombay architectural firm [C.F. Stevens & Company]. Though Bombay was some 40 hours away by train; there was no alternative. All went well at first but in 1910 the man who had been in charge of Institute work retired and the Institute's work began to suffer from neglect during the next 12 months, which was, of course, the most critical period in the history of the plan.

Luckily, Morris had spent some vacation time years ago with his architect brother Wilfred making measured drawings of churches in England so architect's drawings presented no mystery to him. When the meaning of the Institute architects was obscure, Morris and Miller were able to make sense of them and make any necessary decisions regarding construction. No serious mistakes were made; but Morris was ultimately criticized since he often fixed rates for construction work, which should have been referred to an architect, but he had little, if any choice, in the matter. He after all had the objective of opening the Institute by July 1911 and he considered his action an acceptable risk.

It was also one of Morris's expectations that the installation of gas, water and electrical services would have been taken care of by the Mysore Public Works Department. But, all this department did was to instruct Morris to lay a 4 inch pipe line from the main conveying water from a source twenty miles distant to the town filter beds to a tank, which was to be built 500 feet away on the Institute's site. The Institute would have to install a filtering and pumping plant. When the pipe was laid and the tank built, it was found that the pressure in the main was insufficient to force the water through it. An oil engine and pump had to be installed before any water was available.

Then the plans were changed which allowed the Institute to take filtered water from the main, from their water works to Bangalore. This meant working out a scheme with an entirely new pipe line and a reservoir for filtered water on the Institute grounds; by that time it was January 1911, about six months from the date fixed for the opening of the Institute. However, Morris and Miller worked this scheme out and put it through though it delayed the opening of the Institute by two weeks. Even then, water was only available in the laboratories and student quarters the day before the students arrived.

There were other assorted obstacles that Morris encountered. They included losing a shipment of steel roof trusses for the laboratories that were lost in a storm while being transported down the Hugli River and, due to a London dock strike, there was no correlation between bills of lading and vessel cargoes; therefore it was necessary to employ a man in Madras to find their goods from mixed cargoes.

Conclusion of 1910—Start of 1911

Morris and Dorothy had a large and jolly party at their home in Bangalore and the next day Morris went to Calcutta on his own; Dorothy would not leave their daughter alone. Morris went to Calcutta for the December 15th meeting of the Board of Scientific Advice for India.

While in Calcutta, Morris also paid a visit to the Home Department and found his old friend Archdale Earle acting as Secretary. The staff of the home department, many of whom knew Morris well, told him that they were amused at the way he had dealt with Institute affairs earlier in 1910; avoiding disaster for the Institute by taking out a loan at his own risk. Morris also found out here that the home department had nominated him as a prospective guest of the government at the Coronation

Durbar for King George V that would occur on December 12th, 1911. Morris was deeply honored at this invitation given that there was a very powerful move to limit invitations to those who were members of the Indian Civil Service; which Morris was not a member of.

With the Institute to soon open [and this prospective invitation], Morris now, more than ever before, wanted government to make his appointment equivalent to that of someone in government service. He wanted the status to be that of a Director General under government holding a place in the Warrant of Precedence, with the privilege of a private entrée to official functions. Morris saw Viceroy Hardinge's Private Secretary, as well as others, and received what he wanted.

Morris and Harcourt Butler saw each other again and Morris put forward his idea of giving the Institute university status and the right to confer Ph.D. and D.Sc. degrees. Butler, who knew little of university organization, referred him to his deputy secretary who agreed wholeheartedly with Morris on this point. In a letter that Morris wrote to Dorothy, he though wondered if the Tatas would not make it a condition that Morris should back the ideas of Padshah.

Morris was in an amusing position. He was still Director of an Institute that did not really exist. He was supposed to be under the control of a council that could theoretically give him help by supplying him information he might be lacking; however none of those members could be of any help at all. He was also at the mercy of the unscrupulous members of the council who would attempt to curry favor by pointing out his mistakes no matter how minor or well-intentioned.

Everyone in Calcutta treated Morris well and as a colleague; but would this support continue when he was a thousand miles away in Bangalore while facing the Tata brothers and the council? Morris was not in the habit of jumping ahead of or of fearing his fellow man. Given this he adopted, as was the custom of his social class, a family crest and motto, which was "Not rashly—but fearlessly". Morris felt this suited his father Williams's character well and he tried to behave accordingly.

While also in Calcutta he had a meeting with Viceroy Lord Charles Hardinge. The two men discussed the Institute and what Morris thought it would become. Morris thought the Viceroy " . . . was friendly, but very ambassadorial." This was an interesting observation on the character of Hardinge; he had been an Ambassador to Russia, declined the opportunity to be Ambassador to America and later served as Ambassador to France.

With his travels complete Morris set out for home and Dorothy and arrived late at night in Bangalore on January 3rd, 1911. He was greeted and taken home by Dorothy who had learned to drive their Vauxhall car.

CHAPTER XIII

Staffing the Institute, Formal Opening and Royalty

Musings and Personal Health

In April and May the temperature rarely exceeded 100 °F and the nights were generally cool with a breeze. Given the moderate temperatures, the two Dorothies daughter stayed with Morris in Bangalore rather than going on to Ootacamund. Besides finishing the buildings and equipment for the Institute, Morris was heavily involved in drawing up rules and regulations for the Institute, putting together a prospectus and corresponding with assorted officials.

Morris also thought, if he had found upon his arrival in India more government departments ready and willing to take off his hands many of the activities that occupied his time between November 1906 and July 1911 [when the Institute opened and admitted its first students], the Institute may have opened up anywhere from 12-18 months earlier than it did in 1911. Perhaps India had benefited from the difficulties that Morris faced; consequently all learned something new from each stage of the labor.

Before Morris had left England for India in October 1906, he had a complete physical examination and was pronounced healthy. While in India, his health continued to improve [his stomach troubles were no more] and other than a short bout of fever in August 1907, his health was excellent

Additional Institute Staff

In 1910 when the council had been formed and the opening of the Institute was just a matter of time, Morris had put before the council a proposal to appoint further staff, including:

    i.   A Professor of Organic Chemistry;

    ii.   An Assistant Professor of General Chemistry and

    iii.   A Librarian.

According to the Institute's regulations, all appointments were made on the recommendation of a committee in London and later confirmed by the Viceroy. Morris had asked Ramsay to form such a committee; and after a short time Ramsay wrote saying that there were a number of candidates for the assistant professor position but not one for the professor position. Morris did not then understand that, though few university chairs in England paid more than £500 per year, and the Indian positions had no pensions nor retirement allowances attached to them, the Institute's offer of £1,000 per year rising to £1,250 a year was insufficient to attract the type of men that Morris wanted.

Hay, when appointed, had no position given the closing of the Cooper's Hill Engineering College he had previously worked at. Morris knew that Rudolf, who was wealthy, did not see his position in the Institute as permanent. After a while, the candidate who emerged for the Professor of Organic Chemistry position was John J. Sudborough who had been in a backwater position as the chair of chemistry in the University College of Aberystwyth. He badly wanted to get out of this position and was 41 years old when hired.

H.E. Watson, who was 25 years old and had previous research experience at University College London and at Cambridge, filled the assistant professor position. He was later to take Morris's place, upon his departure from India in 1914, as the Professor of General Chemistry at the Institute. In 1932, Watson became Ramsay Professor of Chemical Engineering in University College London.

The committee then also recommended C.H.F. Tacchella for the librarian position.

All three appointments were duly made and confirmed by the Viceroy as Patron of the Institute, who was supposed to give the individual the protection, which the individual government servant would normally receive from the head of the department to which he was attached.

All apparatus, chemicals, machinery, laboratory pipes, fittings and taps for the Institute came from Europe and it was impossible for Morris alone to deal with these details. Rudolf had considerable experience in dealing with imports and was invaluable to Morris and the Institute in this task. Watson, proved his worth early, when it was found, at the last minute, that the 3 inch porcelain pipes for connecting the laboratory bench sinks to the catch-pots below were missing; he consequently improvised and used short lengths of bamboo to solve the problem. Furthermore, when the water supply failed a bit of electrical engineering by Hay allowed water to flow into and be stored for use in a holding tank.

Laying of the Cornerstone

Morris now felt that they were approaching the completion of the first stage of his work in India and he was looking forward to the opening of the Institute in July and the admission of its first students. He felt that on the day the Institute opened no students would be present but everything else would be in order. 24 or so students would arrive the next day and would be personally interviewed first by him. They would then meet with the appropriate discipline's staff member responsible for their work and then meet with the Steward who would assign them their quarters in the hostel. This would create the least delay and quickly have all at work.

Initially, Morris thought an opening ceremony with its speeches would add confusion. But after reflection, Morris thought that the central building (which was to house the general offices and the library and of which only the foundations and plinths were yet built) invited the laying of a cornerstone and that His Highness the Maharajah would be the perfect person to lend his grace, tact and interest to the ceremony.

Morris then spoke to the Maharajah about his idea. His Highness, excited about the idea, consulted his religious advisers who finally decided that February 1st would be an auspicious day for the ceremony. This day was also quite practical given that it was between the cold and hot seasons of Bangalore and so it suited the Maharajah himself, the British Resident, a large number of civil and military officers and many important people including Sir Dorabji Tata who had been knighted during the previous year. Invitations had to be sent out to members of governing bodies, who were many. Morris also wanted "a good show" for the laying of the corner stone.

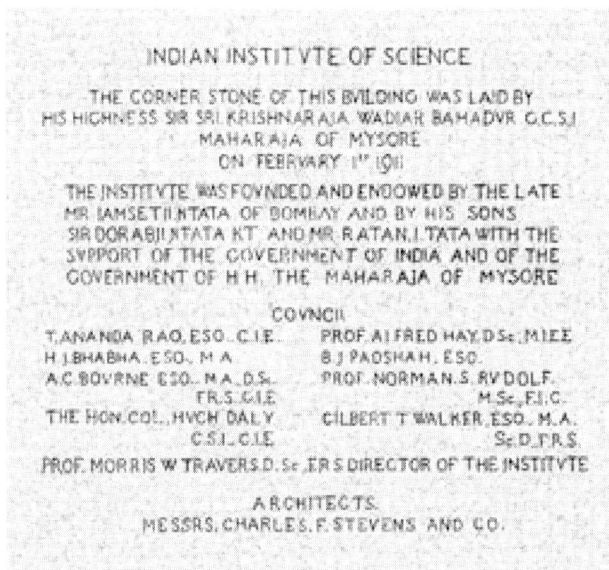

There would, of course, be a large crowd and Bangalore Society would much look forward to the party that Morris and Dorothy would throw at their home that evening.

Sir Dorabji Tata

Morris wanted to give Sir Dorabji the, first opportunity, so far as he knew, to speak in public about the Institute. Morris knew that Sir Dorabji was dissatisfied with what was being done to bring the Institute into being but he hoped that Sir Dorabji would make it clear that he realized all concerned were now making every effort for his father's great experiment to succeed.

He opened his speech by talking of the endowment of the Institute, which he referred to as of unprecedented munificence, showing his complete ignorance of university endowments in the West, and giving an impression totally different from that his father had wished to give, which was that he could do no more than give a lead to wealthy men in India, who, he hoped, would assist in raising the funding needed to establish the Institute. J.N. Tata had known that he could do no more than make a beginning for the Institute.

Sir Dorabji spoke of the Bombay committee of 1898 [though he neglected to say he was not a member of this body], the presentation it made to Lord Curzon in 1899 and of the referring by him of their proposals to the Simla Conference of High Officials and Heads of Government Scientific and Technical Departments; and to their recommendations that an expert should be invited to visit India and report on the plan. He then noted Ramsay's visit to India and his advice to start with a simpler scheme than that proposed by the Bombay committee, which had been devised by Padshah. Ramsay advised that there should be three departments: experimental chemistry; experimental physics and mechanical science and experimental biology.

Sir Dorabji continued in an annoyed manner by saying:

> "Certain gentlemen then drew up a revised scheme, and submitted their report in December 1901, recommending that the Institute should be called the Indian Institute of Science, and devoted to experimental science—with three schools, one of chemistry, one of experimental physics, and one of experimental biology . . ."

He though did distort the facts significantly. These "certain gentlemen' were Professor Orme Masson of Melbourne and Lt. Colonel John Clibborn, Principal of the Rurkee Engineering College. They had been appointed by the government of India with the hope that they might modify the Ramsay scheme so as to eliminate its extension to the teaching of industries, and industrial and commercial activities, which they did, producing a plan that government approved.

Sir Dorabji also left out other very important parts of the story. He neglected to say that his father decided to take personal charge of the development of the Institute and that he realized what he was doing was for the good of India and not himself. In putting this together, J.N. Tata and his advisers recommended the complete rejection of the Bombay committee's (Padshah's) scheme and the adoption of the Masson and Clibborn scheme. The final conclusions that were issued, came from the Bombay committee and included:

> A letter dated April 8th, 1904, and signed by B.J. Padshah, Honorary Secretary of the Bombay Provisional Committee (1898) covering

> i.   A summary of the Masson and Clibborn Report,
> ii. A Memorandum on the proposals by Sir C.J. Martin, forwarded from London.

It was ordered that these papers should be sent to all those interested in J.N. Tata's benefaction. The papers were shown to Morris when Ramsay asked him to allow him to put Morris's name before the committee of the Royal Society that was selecting candidates for the position of Director. Each of these papers was signed—'C.J. Martin' *and the letter carried the signature of 'B.J. Padshah'.*

Morris now realized that Sir Dorabji's personal knowledge of the story was only up to the Ramsay report. It was clear that Padshah had induced him to believe that J.N. Tata's last wishes were embodied in the Bombay provisional committee report of 1898, which Padshah had written.

The final part of Sir Dorabji's speech was a tangled story involving financial details that only Morris knew completely. He referred to the generosity of his family in financing the Institute at the beginning [in fact money due to the Institute from the Tatas did not come to it at all while Morris was its Director from 1906-1914].

Only after February 1st, when Morris had in his hands the press report of Dorabji's speech and had put it in with his files, did he really began to understand the Tata position; but until he better understood Padshah and how he managed to exert such influence over the brothers he never would really understand what was going on.

Education Conference at Allahbad

In mid February 1911, Morris received a personal invitation from Harcourt Butler to attend an education conference at Allahbad. Morris was hesitant to go given that there were likely to be many members of the Indian Education Service present but he did and gave a presentation on the difference between the proposed organization of the Institute and the colleges that existed in India.

In an Indian College, Morris noted that control was entirely in the hands of the Principal who was generally appointed due to his seniority without so much as a board of studies to advise him. A professor was basically one of his assistants; this was the way it had been in Bristol when Morris had arrived there in 1904.

On the other hand, in a modern British university there were independent departments each under the direction of a professor or independent lecturer but interdependent in the respect that professors and independent lecturers were members of the senate who were under the chairmanship of a Director or a Principal. This may seem contradictory but it presupposed that goodwill prevailed and that all in the institution were working for the common good. Important matters to a department were discussed and voted on before they reached the senate or the council. Though the council, which had a minority of professors on it with its majority being non-academic individuals, was the governing body of the institution it did not dominate the institution. Real institutional efficiency depended on very close cooperation between the senate and the council.

Morris, in describing the formation of the Indian Institute of Science, said that with a very limited amount of money available for capital or current expenditures, it was quite difficult to decide which five departments should start the Institute. It was also necessary for the Institute to have independent departments that could cooperate with each other to help in trying to solve the technical and scientific problems of India. Therefore, independence with interdependence was vital for this body.

Morris continued by noting that Ramsay, then Masson and Clibborn and then he had decided that the three great divisions of science and technology that must be represented and would comprise the Institute when it opened in July would be: chemistry (pure and applied); physics and mechanical science and biological science. Morris finished his presentation by noting what had appeared in the press, that what he was engaged in was a great experiment and that this was all due to J.N. Tata, Sir Sheshadri Iyer, the government of India and the Mysore Durbar.

Dasara Festival at Mysore

In October, Morris and Dorothy were guests of the Maharajah of Mysore at Mysore City for a week to celebrate the Dasara festival. At this event, a camp of large two-room tents was set up on the grounds of the former British Residency as well as a banquet hall and recreation rooms.

This festival was a survival of the autumn festival, which had religious associations. During this event, the Maharajah and his court went into retreat for several days. Then, they went to a Temple that the assembled guests were allowed to see. Being still in retreat, the Maharajah and his court rode on elephants in curtained carriages, being almost invisible and dressed very simply. At the Temple they dismounted and spent some time in ceremonial activities. Guests then left, dined and returned after dark to see the return to the palace with the Maharajah wearing his state robes. For the procession the elephants, masquerading as sacred white elephants, were painted pale pink all over and were decorated with conventional designs in gold leaf.

As always on such occasions there was a certain mix of the sublime and ridiculous. The military aspect was provided by a contingent of the very efficient imperial service corps of cavalry commanded by the Maharajah's brother-in-law, Colonel Desaraj Urs, aided by a staff of British officers. But the artillery was limited to a few small brass guns that were perhaps from the very beginning of the 19th century. Morris and Dorothy saw all the events very well and were glad that they were able to attend. They had been invited to the festivities in 1910 but Dorothy Mary was due and they had to decline.

The Coronation Durbar and its Aftermath

After being informed he would be "invited" to the Coronation Durbar while he was in Calcutta in December 1910, Morris never received a formal invitation but received several letters giving him details about the event. Uniform, or court dress would be worn during certain events; for this Morris wore his crimson gown, sans hood, which went well with his suit. Morris and Dorothy were told in these letters that their quarters for the event would be a two-room tent, 26 ft. by 18 ft. They could bring their own car or share an official car (taxi). Morris and a retinue of servants, but not Dorothy, would travel at the expense of the government.

This was an exceptional honor; for as it turned out, the government of India camp was very select, and the guests and their ladies numbered only 110 together and they included members of the Viceroy's Council, the Judges of the High Court and high officers of the government such as the Advocate General. The Director General of the Survey of India, a Colonel Burrard, represented science and Morris represented education.

Morris and Dorothy left Bangalore for the event on November 29th and arrived in Bombay on December 1st, staying with their friends Mr. and Mrs. Burns at the Art School Bungalow; Burns was a common but amusing English lawyer while Mrs. Burns had recently worked for the Bandman opera company. They dined that evening at the homer of Sir Dorabji Tata in the Esplanade. The dinner was quite good, and all had an entertaining time inspecting the art treasures of Sir Dorabji.

The next morning Morris and Dorothy were at the yacht club to watch the King Emperor's arrival. The ships steaming into the harbor took their positions with perfect precision but Dorothy could not tolerate waiting in the sun for the five hour state entry ceremony to be completed.

So they left for Delhi that afternoon by train. They occupied a whole compartment, traveled comfortably, and arrived the next evening. The Delhi station was a howling pandemonium but an officer from the camp met them there and they went to their car (which had been sent to Delhi from Bangalore). Morris then drove them to camp but did not have any thick gloves and had difficulty in handling the steering wheel near the end of the trip.

Morris and Dorothy were provided with one of the large double tents at the corner of "Central Road" with their neighbors being Gillan the Auditor General and Captain Lumsden, R.N., Director of the Royal Indian Marine. In the next line of tents were the judges, and at the opposite end of the camp the members of the Viceroy's Council occupied the larger tents. In the middle of the camp there were large marquees for dining rooms, drawing rooms, a smoking room, billiard rooms, etc. Morris and Dorothy were quite fortunate; they had tickets to attend many functions that other more notable individuals did not.

It was said that the cost of this little "outing" was on the order of £1,000 a head per guest. Whatever the cost, no entertainment was ever better organized, or better arranged in every way. The tents were new while the plates and cutlery, china and glass were made special for this event and bore the government of India Crest. The camp itself was laid out with perfectly made roads and gardens. Cooking and meals were outstanding. Every day one received a program giving the minutest directions regarding such functions as had been arranged. Naturally not all was perfect; polo players did not design the polo grounds and real privacy was not possible for anyone, even the King Emperor and the Queen Empress.

The first important function that Morris and Dorothy attended was the state entry of their Majesties. A circular amphitheater had been constructed, with covered and raised seats in two semi-circles, the arcs of the circles being filled with lawns on either side of the processional road, which formed its diameter. These seats were for the mighty, but for some reason or another it was decided that the principal guests should be seated on chairs on one of the semi-circular lawns, and the covered seats in the semi-circles should be given to the less important. The amphitheater was actually on the summit of the ridge, to the west of which lay the camps of their Majesties and the government of India. On the lawn to the east of the processional road were seats for the members of the council and other very high officials. Less important, but yet fortunate, individuals including High Court Judges as well as Morris and Dorothy occupied the lawn on the opposite side. Morning dress or uniform was prescribed and though it was not warm in the shade, the sun was fierce.

All had to be in their places very early, and the waiting time was spent by chatting with friends and laughing at the attempts inappropriately dressed individuals made to avoid sunstroke. After this waiting period was done, then came the procession of the ruling princes of India, which was a pageant of the diversity of the Indian Empire. Every ruler, no matter how much land he held, and who could possibly get to Delhi, was there to honor the King Emperor. Each of these rulers was doubtless spending all or more than he could afford and had brought with him all that he possessed in trappings and finery.

The Coronation Durbar on December 12th was the principal function. Everyone dressed in their best; Morris in his scarlet doctor's robes over court dress while Dorothy wore a very pretty violet dress with hat to match.

The ruling chiefs under cover occupied the front row of seats and behind them were High Officers of the State. The guests in the government of India camp were accommodated in the next four rows and twelve places on either side of the centerline. Morris and Dorothy, being the juniors in the camp, were the 11th and 12th from the center in the 4th row from the front, which brought them onto the exact level of the durbar pavilion and in line with the posts supporting the canopy. Thus, they were a scant few feet from the King Emperor and his Queen Empress. They could see the ceremony slightly from one side.

Under the canopy of gold and crimson, were seated the King Emperor and Queen Empress in their Coronation robes and crowns, the jewels and gold flashing in the brilliant sunshine. Around the throne were grouped the Viceroy, the Secretary of State for India, and the members of the court and at the foot of it were seated young Indian Princes in brightly colored Indian costumes. All around was color and glitter and in the distance the uniform of the attendant troops and the large number of Indian spectators formed the background to the brilliant scene.

The King Emperor first received the homage of the rulers of India. Each ruler advanced, bowed to his Majesty, or saluted and thrust forward the hilt of his saber to be touched. He then bowed to the Queen Empress and then retired. After the Nizam paid his homage, Baroda created a giant stir when he strolled up to his Majesty, tapped his stick on the ground, made something between a nod and a bow, ignored the Queen Empress and turned his back sharply as he walked away. The spectators gasped and it seemed to Morris that Baroda's intent was to insult.

It was though the case that Baroda had no proper appreciation of his duty to the King Emperor or of their relative positions. One had to remember that he was not born to rule and had been brought up as a child in a small village and later adopted as the heir to the Baroda name. Though he was likely a keen and able ruler, and an adept businessman, he hated ceremony of all sorts and would not "dress up" to please anyone. He even said to Lady Fraser on the morning of the Durbar that he did not see why the Indian princes had to do homage in public. As a result of his experience with the King Emperor, he left Delhi in disgrace.

After the magnificence of the Durbar ceremony, everything else was a bit disappointing and most of the guests simply wanted to go home. Morris and Dorothy went to the reception that evening and to the review of the troops, which was a grand sight and the investiture the day after. The state departure was more imposing than the arrival of the King Emperor and the Queen Empress. The King Emperor wisely rode in the carriage with the Queen Empress. Morris and Dorothy saw the procession pass from a position on the ridge, which though was not full. Soon, they left for home in Bangalore by rail.

CHAPTER XIV

Affairs of the Institute

Official Jealousy

Morris now did experience what was the greatest obstacle to progress in India; official jealousy. In India every official belonged to a "service" with that service giving the individual prestige or bitterness and envy if it was not one of the 'best'. The designation "I.C.S. (Indian Civil Service) that a man could write after his name put him in a class above all others. Promotions were generally within each service and only, on rare occasion, might individuals be brought in at higher levels from outside.

Morris had been properly warned early in his Indian career that it was considered that his appointment should have been given to an Indian official, ideally a member of the Indian Education Service (I.E.S.). However, at the beginning of his time in India his early contacts had been with men in high-level positions who had no reason to be jealous of him. He also felt that he was outside the inner circle since he also dealt with such "small fry" as directors general.

He had learned enough about internal official jealousies during his first year in India to decide to refuse, if at all possible, to be a member of all official boards and committees dealing with university matters; and when the Governor of Bombay asked him to advise on the reorganization of the university, Morris agreed to do so only on the condition that he do so in a private capacity and that his name was not be mentioned officially. When the Governor of Madras wrote asking him to attend an education conference in Ootacamund, Morris felt duty bound to accept the invitation. Morris, who was treated by the Governor [whose action was later strongly disapproved of by the Madras inner social circle] with exceptional honor, later regretted this socially.

Morris was much aware of this isolation. In February 1911, when he accepted the invitation of Sir Harcourt Butler to attend a conference at Allahabad, where a large part of the audience were I.E.S. members, he intentionally tempered his comments by not saying what he actually thought of Indian university education but did little more than answer questions about the Institute when they arose.

Another example of this isolation was in the very cold and formal reception Morris [and Dorothy] received when they called on the new Resident, Sir Hugh Daly in Mysore, the chairman of the Institute's council, after the Coronation Durbar. The Resident, who had only received a K.C.I.E. honor rather than a K.C.S.I. honor, was personally hostile to Morris who had after all been the *only* individual associated with education invited to the Durbar. This hostility was to be apparent soon enough with respect to the Institute and would become a very big problem for Morris.

Affairs of the Institute

In July 1911, Morris had received a letter from the education department of the government of India asking him to

> "draw up a scheme for the development of the Institute as a complete University Faculty of Pure and Applied Science, with an estimate of the cost and of the number of years over which it would be spread".

Morris, who was at that time very busy with the opening of the Institute, could not reply before he visited Delhi. He, upon reflection, felt that this was an impossible request. He thought he could produce a scheme for an institution modeled on the faculties of science and technology in a modern Western university; but he was at the beginning of carrying out a difficult experiment aimed at seeing what type of institution was best for the needs of India.

He instead preferred to make an analysis of the problems that he had to face, which came under four headings:

    i.   Relating to the subjects to be dealt with, and the order in which new departments were to be created;
    ii. Relating to management;
    iii. Relating to staff and
    iv. Relating to students.

He had already dealt with the problems of immediate development of the Institute in his 70-page pamphlet, which he had written for the completed council in 1909 and accepted in July 1910. This pamphlet involved the establishment of departments representing:

    a.   Chemistry;

b.   Mathematical and physical science and

c.   Biological science.

The ultimate object of this plan would be to establish interdependent, and yet independent departments within. A department of archaeology was out of place in a group of departments dealing with chemistry, physics and engineering. The chemistry group already had three departments with three professors and one assistant professor. The Director of the Institute had personal charge of the department of general chemistry and it was expected that he would find time to engage in research. This was only possible though if the Institute was small and if the director was already a keen researcher. A larger Institute would of course "free" the director from departmental administration. Morris considered that with professors [and assistant professors] in general, organic and applied chemistry, the experiment that they were engaged in might be tried out in the case of chemistry. He proposed to strengthen the department of applied chemistry by appointing an assistant professor to concentrate on chemical engineering. Specialties such as analytical and agricultural chemistry could be added as time and resources permitted.

The second group covered mathematical physical science and mechanical science. Though isolated departments in mathematics and physics might have attracted a few brilliant students, this would have overlooked the wishes of J.N. Tata who wanted science brought directly to the aid of Indian arts and industries. Morris therefore advised making a department of electrical engineering the nucleus of this group. Students for it were at once available from the Indian engineering colleges and there was a significant demand for trained electrical engineers. Morris now proposed to advise the council to immediately appoint assistant professors in this department to provide instruction in applied mechanics and in applied mathematics. Following that, a chair of physics and mechanical engineering would need to be appointed.

Morris had also, in his previous reports to the council, recommended the appointment of a professor of applied bacteriology whose department would form the nucleus of the biological group. Next would probably be the appointment of a professor of economic botany though the extension of the Institute into biology must involve close cooperation with the Mysore Department of agriculture; Morris had had discussions about this already with the Mysore Durbar.

Morris summed the matter up by saying that with the increased staffing of the four existing departments, with increased provision for experimental work, and the formation, in order, of departments of applied bacteriology, economic botany, mechanical engineering and physics, the experiment that was being conducted could then be fully attempted. The annual cost would rise from Rs.270,000 to Rs.500,000 and the capital cost from an estimated Rs.1,900,000 to Rs.2,800,000 of which about two-thirds was available.

Morris had, in conversations in Delhi and in a letter, noted to Sir Harcourt Butler that effective management of the Institute by its present council was impossible since the council did not work through committees with executive authority. Also, there should have been a small executive committee of individuals living in or near Bangalore as was suggested in the Masson and Clibborn report; but the council would not do this nor even appoint a vitally important buildings committee.

It was also difficult to obtain a suitable staff. Only a single candidate had offered himself for each of the currently filled chairs. Though the starting salary offered was about twice that what a professor would receive in England (£1000 vs. £ 500), difficulties arose from the fact that the retirement age from service in India was 55 while from British service it was 65 and the limited income of the Institute made it possible to offer only retiring allowances calculated as a percentage on salaries earned, instead of pensions, as was so in the Indian Education Service.

At the beginning of the Institute, it was of course difficult to attract sufficiently well trained students to take proper advantage of study and research. Only 24 students had been admitted and retained during the Institute's first term and less than half reached the hoped for academic standard. However, on a positive note, Morris's advice that he had given on the reorganization of science teaching both in Bengal and Bombay had been heard and preliminary results were promising.

Council Meeting of February 1912

After the Coronation Durbar, and as 1912 began, Morris put together these and other ideas in a report that was then sent to all members of the council. His ideas came up for discussion at the council meeting of February 7th, 1912. In this report, he referred to his communications with Simla and his conversation with Sir Harcourt Butler in Delhi.

The Resident, as the chair, then said

> "I think that we should pass a vote of censure on the Director for corresponding with government without our permission."

Morris at this point could not even hint at the fact that he had come to India to advise government about the Institute and that it was *his views* and not those of the council that had been sought. Very angry, he said nothing and waited. His anger soon turned to amusement and hatred, which was what he usually felt from council meetings.

The room itself was totally silent. Sensing that the chair had made a grave error, Morris then read a resolution which the council was aware of stating—"That it is the immediate policy of the Council to develop the Institute as a Faculty of Pure and

Applied Science." Morris explained this resolution by noting that the provisional committee of 1908 and the incomplete council in 1909 had the identical resolution; he felt that the complete council should now approve this.

A Tata representative noted he was authorized by Sir Dorabji Tata to tell the council that Morris's resolution was beyond the powers of the Institute and the council. If the council passed this resolution, the Tata brothers would take legal action. Morris said nothing and the council passed his resolution. Further discussion on this continued all morning and the better part of the afternoon and ended with the passage of a resolution stating

> "That the Council was of opinion that the only work of the near future should be the fuller development of the work of the existing departments."

Academic members of the council declined to vote on this matter, opining that the resolution had nothing to do with what was being discussed. Morris here also believed that the Tatas would not carry out their threat of legal action because they would have to establish their right to interfere in the affairs of the Institute. This claim would be seen to have no legal basis to it.

Just before the council adjourned for lunch, an incident occurred that would have been impossible under Stuart Fraser's chairmanship. Morris overheard a whispered question to the Resident—"Are you going to bring that matter up now?" The Resident then told the council that Morris had not only admitted to the Institute a student with a very bad political record, but had made him a demonstrator. Morris then had his secretary get X's (who was one of Morris's two first-rate students) confidential file, which he then read from:

> "About four years earlier Mr. X. had been a student in a College, better described as a secondary school, in a native state, in which the Chairman had then been Resident. Mr. X. had attended a political meeting. This had come to the knowledge of the British Resident who had ordered the boy to be expelled from the College, which he had done, in spite of a protest by the Principal. He then went to Presidency College, Calcutta, where he had spent three years, coming to the Institute with an unimpeachable record, politically and educationally."

This file had, as usual, been sent to the Resident who had *so conveniently* forgotten about X. There was just nothing more to be said; and not even the D.P.I.'s remark—"I suppose that the Director has broken some Bye-law,"—could continue the discussion. This discussion was resumed when Morris was away in July 1912 and X's demonstratorship was terminated.

Morris now proposed to take three months leave as well as the long vacation so he would be away from April to mid-October. He had reservations about leaving for so long especially after reflecting on this February 7th meeting. He actually wished to wait a year to take leave but he did not want to interfere with Hay's leave arrangements for 1913.

Mid February and March—1912

The quality of most of the students who had been admitted and retained in the Institute was better than Morris had expected though not what he had hoped for. He believed that most of them would benefit by spending a couple of years [or longer] at the Institute. A few of the exceptionally qualified were engaged in research. All the students were also satisfied with their living arrangements. The staff appeared to be happy and contented. Morris hoped that at the March 18th council meeting he would be able to further explain the Institute and its needs. He had been fortunate when Stuart Fraser was the Resident and Madhava Rao was the Dewan because they took a real interest in the Institute and its progress in general.

However, Morris had now encountered a Resident and Dewan who were polar opposites of Fraser and Rao. Even though the new Resident and Dewan were the only non-academic members of the council who attended council meetings regularly, the Resident regarded the Institute as a possible center for subversive political activity, in Morris's opinion. He was far from being a well man, seeming to suffer commonly from memory loss. It was also stated that he had been given this position as a sinecure.

The Dewan was always friendly and courteous but Morris thought that the Dewan regarded him as radical. One of Morris's few personal contacts with him related to the reinforcing of road culverts to enable brick suppliers to use a steam-propelled lorry, an innovation, to transport bricks to the Institute. The Dewan thought it was amusing that Morris believed motor transport would be an important part of India's economy in the very near future.

Morris, before going on leave, tried again to give the council at its March 18th meeting a good picture of where the Institute was and where it was going. This effort was handicapped by the fact this council was the third governing body which had existed since January 1908.

There had also been two periods when the temporary governing body had been deprived of authority. During these periods Morris had the choice of canceling building contracts or of carrying on. Morris decided to carry on. When they got to work after the second hold-up in 1910, Morris had to organize matters so that it would be possible to open the Institute in July 1911. This was only possible by stopping all other work and making his building staff devote all their time to supervising the complex work involved in building and equipping the laboratories.

The result was that bills for buildings and equipment were in arrears and Morris had to arrange with the finance committee to make advances to the contractor for amounts certified by the architects and him. The other two members of the finance committee, the Dewan and Professor Hay, would be at the council meetings while Morris was away.

The building staff could finish measuring and making out bills in Morris's absence. New work could then be started. Professor Rudolf (who would act as Director while Morris was gone) asked for an extension to his laboratory and for some new staff bungalows. Morris asked that plans and estimates should be sent forward on these projects while he was on leave/vacation.

Morris asked the council at this March 18th meeting to sanction appointments of two assistant professors in Hay's department to deal with applied mathematics and applied mechanics and one in Rudolf's department to deal with chemical engineering. In addition, Morris offered, unofficially, to assist in looking for the men for Hay's department while on leave in England.

CHAPTER XV

Personal Matters, Tatas, the Auditor's Report and the Council

Personal Matters

Dorothy, nurse and baby sailed for England from Madras in late February 1912 while Morris was busy until the end of March 1912 when he followed them, traveling via Madras, Tuticorin, and then by sea to Colombo. Then, anxious to get to England, he went through Port Said and Brindisi before arriving in London.

Three days after getting to London, Morris went into the hospital for treatment of a hernia. While in the hospital for three weeks, Morris learned that Dorothy had to send their Eurasian nurse back to India at their expense and had taken charge of little Dorothy's care herself. When Morris got out of the hospital, they took a fishing vacation first at Eggesford and then at Simonsbath on Exmoor where Morris's mother joined them.

Work followed Morris even on his vacation. He was in correspondence relating to the appointment of the two assistant professors in Hay's department. Morris had found two first-rate youngsters willing to go to India on terms he had laid out. Later, Morris interviewed both of them and sent their papers and his views about them to Hay telling Hay that the final decision rested with him. Morris told the two young men that they would hear more from India about their possible appointments. Hay acknowledged Morris's letter and said that he accepted Morris's recommendations and would follow the Institute by-laws for hiring new staff.

After the fishing trip, they returned to London early in June and were quite happy. They had, at Anne's home, a wonderful dinner for the Ramsays and Sir Rajendra Nath Mookerjee, head of the engineering firm Martin & Co. Ltd. of Calcutta, and his wife. Morris also had the opportunity to represent the Institute at the Royal Society's 250th anniversary meeting in June and was able to shake hands with the King Emperor and the Queen Empress. He also attended the Congress of the Universities of the Empire and had conversations with many who were facing the same problems he was dealing with. Morris found it was generally agreed that postgraduate institutions such as the Institute should be able to confer the Ph.D. on students who engaged in research.

Morris and Dorothy hired an Irish nurse for their daughter and the four of them went to Norway in early August. They spent most of their time at Sandene on the Nord Fjord and at Red on a lake a few miles inland from the fjord. Morris did not succeed in making Dorothy into a fishing enthusiast but they enjoyed the vacation anyway. Morris ended up catching a fair number of fish and finished up with his greatest success ever in fishing; catching a dozen trout all weighing in between 2-3 pounds at the lower end of the lake in a short length of river.

The Tatas

Morris was beginning to find that after three weeks of vacation he was ready to get back to work. He already knew that trouble was brewing for him in India and some of this came in a personal letter he received in early September from Sir Dorabji. First, Sir Dorabji proposed to erect a statue to his father and have it located on Institute grounds. Morris opposed this idea for two reasons: first that it was entirely contrary to what J.N. Tata wanted and secondly that Sir Dorabji only offered to pay for 25% of its cost. The second part of the letter is reproduced below:

London
September 9th, 1912

Dear Dr. Travers,

I have to acknowledge your letter of the 28th August from Norway. I saw Gilbert Bayes yesterday and he has made several sketches for the Bangalore Memorial which seems to be very good. But in the absence of the plans from Bombay it is difficult to finally decide. Still he was able to see Gregson, one of the firm of Stephens & Co. Bombay, who gave him a rough idea of the ground plan of the whole plot and of the elevation of the main building, and he has worked up to it. There are many things in this connection which it would have been very much better if you and I could have met and talked things over with Bayes so as to have a scheme which both of us could approve of, as it would have saved a good deal of time in correspondence from long distances when exchange of view is very difficult, if not impossible. We are leaving for Baden-Baden to do my usual "cure" on the 11th and will not be returning to London till the last week in October, after two weeks travel in Germany after the "cure". You will arrive in London after we have left and leave before we arrive. This is very unfortunate.

There is another matter on which I take this opportunity on writing to you. You know that at the time my father conceived the first idea of the Research Institute at Lord Reay's instance his intention was to make it specially an Institute where medical research would be the main feature. Lord Curzon's attitude and Sir William Ramsay's report somewhat diverted the original aim, but it has always been before me. Realising that the present scope of the Institute is not wide enough to cope with the idea of a School of Research in Tropical medicine which I am anxious to see established in India, I propose to set aside funds or properties which will bring in something like (£5000) a year and devote a part of it towards the establishment of such a school, and the rest to one or two objects for the benefit of India which also I have long had in mind. With this view I put myself into communication with Major Rogers in India to ascertain if the remaining two chairs of the Institute could be made available for the purpose, and what additional endowment was necessary for a workable scheme that could be effective.

If the Council of the Institute will come to my aid and devote the surplus of its present income towards my scheme, I am prepared to endow the Institute further to such extent as will make it possible to do useful work in the direction of Research in Tropical Medicine, with the aid of the Government of India's further contribution. But if my views do not meet with support in this direction then I may devote the intended amount to endow the school now proposed to be established in Bombay or found chairs of Research in connection with other schools in Calcutta or in England, and then devote a larger part of the sum originally ear-marked for medical research to the other purposes I have in view.

My first aim is to try and make the Institute more important than it is, but if I find difficulties in doing so, that is if I find that the amount I propose to devote to it is not sufficient without the co-operation of the Institute, I shall try to make my other scheme the main scheme and the Research in Medicine only subsidiary.

I am posting to you under another cover a copy of Roger's note which I have printed. I may say that I have already consulted Prof. Nuttall of Cambridge, Sir Clifford Albutt, Sir John Rose Bradford, and at Lord Crewe's suggestion, Sir Havelock Charles, and the later as the medical adviser of the India Office has promised his full sympathetic help in furthering the progress of Medical Research in India. If the Council of the Institute seconds my efforts, as Rogers points out, Bangalore could be the centre round which all other Medical Research would be carried out in India. I am having copies of the Pamphlet distributed to all members of the Council and Standing Committee for their views, and I shall be much obliged if you too will do so.

Yours sincerely,

D.J. Tata

The day after he had received this letter Morris spoke with Sir Havelock Charles. Charles read the letter and burst out laughing. He said that Sir Dorabji had spoken to him in a general way; but he had never mentioned Bangalore, where there were no hospital facilities except at a small military station hospital staffed by army medical corps officers. Charles said that if Sir Dorabji really wanted to promote medical research in India, his professors would need hospital facilities in the large cities; so the idea proposed was ridiculous.

Morris then wrote back to Sir Dorabji informing him of his conversation with Sir Havelock Charles and that in Morris's opinion Bangalore did not have the facilities needed for Sir Dorabji's idea. The proposal could not be put before the council. In reply, Morris received the following from Sir Dorabji dated September 23rd, 1912:

"I have never understood your point of view and do not understand it now. All I know is that my brother and I gave a certain amount of money to carry out certain aims which my late father had in view. You were appointed director and you thought the money should be spent in the way you thought best and that it was enough if we had found the money, the rest did not concern us. Wiser heads than ours who understood the needs of India better than we did (owing to two or three years residence in the country) were the proper people to decide what was to be done with it. I remember, when we were discussing these things, my telling you that though my brother and I had given the money we thought that it should be spent for the good of India. I added that I considered that the money was in trust for the people of India and that we were the Trustees. I distinctly remember your answer word for word. It was—'Don't talk to me in copy-book maxims'.

Some years later my brother offered to endow a chair for a special purpose but you would have nothing to do with it. Now I propose to add to the funds of the Institute for a purpose avowedly of benefit to India and which must add dignity and influence to the Institute and to my surprise you reply that you must refuse to support my proposals coupled wit my conditions. Even in the matter of the employment of the income of the property handed over to the Institute, we as donors insisted that the income should not be employed on brick and mortar but only for the purposes of teaching as it was only for that purpose that we had given it. But owing to the majority you controlled on the Council you over-ruled this and spent the money for purposes for which it was not given on the ground that after it was given the Council had the best right to spend it as it chose to. This majority was introduced into the constitution entirely against my will. I still maintain that the right course is for professors to teach, and not to concern themselves with internal management."

Every paragraph in this letter, in Morris's opinion, was a lie. The Tata brothers were the executors of their father's Will and nothing more. There was an agreement between the government of India, the Mysore Durbar and the late J.N. Tata. The terms of Morris's appointment were clear and definite.

Morris felt Padshah had deliberately misled him as to the wishes of the Tata brothers. Morris never denied comments he made nor did he advise the council to reject Ratan Tata's offer. Morris also thought the statement absurd that he controlled the council; it had always been Padshah, sometimes supported by Bhabha, against the rest of the body.

Morris sent copies of his correspondence with Sir Dorabji to members of the Court of Visitors. Of the fifty members of the court, seven supported Morris, one supported Sir Dorabji and the rest did not reply.

Report of the Auditor

Near the end of August 1912, while he was still on his fishing vacation in Norway, Morris received a letter from a friend in Bangalore who had no connection with the Institute. This note, complete with a cutting from a local Bangalore newspaper, told of the summary dismissal of Miller, the engineer in charge of building construction of the Institute. No reason was given for this action; the editor of the Bangalore paper described this action as arbitrary and unjust.

While Morris was in London he received on September 23rd an official letter from Rudolf dated September 4th that contained a copy of the Auditor's report for 1911-1912 and a copy of the minutes from the council meeting of August 27th, 1912.

When Morris began looking at the report, he immediately noticed something wrong. It opened with a statement that the account for assets and liabilities showed a deficit of Rs.500,000. Though he had no papers with him, Morris's memory for figures was good and he quickly realized that the auditor had charged every project Morris had put before the council as a liability. What was real was the exact opposite of that "revealed" by the auditor's report. It was clear to Morris how this "error" had arisen but what puzzled him was—Why did Rudolf and the council not detected it? It certainly began, in Morris's mind, to look like a "frame up".

Morris then read the first few pages of the auditor's report. It was quite clear to him that an extraordinary mistake had been made and then he turned to the typed paper accompanying Rudolf's letter. The meaning of a second letter he had received from his friend in Bangalore was crystal clear—the author or authors of it hoped that Morris would not return to India.

But he or they were sadly mistaken. Morris's only concern now was to get to grips with whatever problem this foolishness involved. He was not at all concerned for himself. It though made the work of building the Institute, which was now finally starting to succeed, much more difficult. Morris was convinced that the Tata representatives and government officials backed at least by the Resident and the D.G.I. were behind all of this sordid mess. The conduct of his own colleagues puzzled and worried Morris though; he believed that they must have felt hopeless to try and back him against so many foes so he would not conduct a witch-hunt when he returned.

Reaction of the Council

The council, in its meeting of August 27th, 1912 reacted by stating in two resolutions that Morris had kept the council in the dark about the financial position of the Institute. A third resolution, and its parts, was:

> "The Council having read the Report on the Audit of the Accounts of the Institute for the year 1911-1912 resolve:

> i. That the Director, Dr. M.W. Travers, be asked to furnish the Council with such explanations as he wishes to submit regarding the irregularities which have been brought to light.

> The Council consider that Dr. Travers as Chief Executive Officer of the Institute has gravely failed on his duty to the Council inasmuch as he has neglected to inform them of the true state of affairs in respect of the progress of expenditure on the building works; that he has usurped the powers of the Council in authorising additions, alterations, and deviations from plans without either obtaining the permission of the Council, or informing them of what he had done; and that he has committed the serious error in making large advances to the Contractor, and neglecting to observe the terms of the contract relating to payments.

> ii. That Dr. Travers's attention be called to a statement by the Engineer as to the excess of costs over estimates. The Council deserves an explanation of the cause of excess in each case.

iii. Dr. Travers be requested to explain why he took no steps to ensure that measurements by the Engineer were not checked measured by the Architects.

This last resolution was followed by a series of further resolutions that were so absurd that Morris laughed when he read them. He was only willing to comment on two of them, with the first one reading:

"That contrary to the usual practice the Director obtained the permission of the Council for granting local contracts for the supply of bricks instead of relying upon the general contractors to supply them, whereby affording the Engineer the opportunity of obtaining illegal gratification."

The buildings of the Institute were originally to be stone. Morris, who had the idea of using brick for partition walls, knew that brick was advantageous, as did the provisional committee when they approved its use in 1908; two members of this body had been the Resident (Stuart Fraser) and the Dewan (Madhava Rao). Further, at the council meeting of July 14th, 1910 the officiating D.I., who was an engineer, had praised the use of machine made bricks as an unusual economy.

The second resolution of note dealt with the appointment of the two assistant professors for Hay. Morris had found two men he considered very suitable, and had sent the facts regarding them to Hay, telling them that they would be hearing from India, and asking Hay to expedite the acceptance of them as they would be applying for other positions. But Hay did not even write to them [thereby making a terrible error], so one of them supposing that he had been appointed, after waiting two months, began his travel to India. Therefore, there was a resolution (which Hay voted in favor of!), accusing Morris of:

"Taking upon himself

i.   The functions of the Senate as an advisory body;
ii. The duties of the Council as the appointing authority and
iii. The privilege of His Excellency the Patron.

Morris, in commenting on his 'crime' of "advances to the contractor", that every check for more than Rs.100 was signed by the Dewan, Professor Hay and Morris at meetings of the finance committee and very full records were kept of this committee's meetings. Amazingly, no reference was made to these two key facts and amazingly, both the Dewan and Professor Hay supported this questionable resolution.

To Morris it was clear that he could form no picture of the background of what had happened. However, he was extremely surprised that the Resident in Mysore, Sir Hugh Daly, had committed himself to what was certainly questionable if not libelous conduct.

Morris began thinking of what he would do upon his return to India.

Chapter XVI

Personal Attacks, Special Committee and Student Grievances

Return to India

On the voyage back, Morris was fortunate to have a single-berth cabin and he had the time to put down some thoughts on paper on the charges of alleged misconduct. He also drew up a petition asking the Patron to appoint a committee to look into not only these accusations but into his administration of the Institute and the problems he had to face over the last six years.

When he arrived in Bombay, Morris had a wire from Dorothy telling him that the papers that had been sent to him in London had also been sent to Simla. So Morris went directly to Simla from Bombay and handed in his thoughts on the misconduct issue and his request for a committee to His Excellency the Patron on October 15th, 1912; with a covering note sent to Sir Harcourt Butler. Morris received a reply back from Butler, in a very friendly tone, that since he would have to adjudicate on the case he could not discuss it with Morris at this time. Morris was also advised to return to Bangalore and keep quiet while the case was being considered.

Morris talked with members of the staff of the Department of Education that he knew and they told him that no importance had been attached to the statements that the cost of buildings and equipment were in excess of the estimates. They were of the opinion that a personal attack by a member of his own staff was being waged against Morris.

Morris returned to Bangalore directly from Simla, arriving on October 26th. On that morning he had breakfast with his friend Frank Usher, from the Central College, and drove to the Institute. He arrived early and saw a stranger where he expected to see his chief clerk and friend, Sundaram Iyer. Iyer had been fired!

Rudolf then arrived and Morris took back charge of the Institute from him. Their conversation began by Morris asking Rudolf what exactly had happened in the last two months. Though Morris wanted to try and re-establish relations as they

161

had existed before he went on leave, he was very cross with Rudolf. Rudolf told Morris that he had acted while Morris was away on instructions from the Resident Sir Hugh Daly who was chairman of the council. Morris now obtained printed copies of the minutes of council meetings and was amazed to find out that no minutes of meetings of the buildings committee [which had become the de facto executive committee of the council] had been kept.

While in Norway, Morris had heard of the firing of Miller. There was no reason on record for this action; but Morris later found several acrimonious letters where it was stated that Miller had taken bribes while Rudolf vehemently noted that the so-called evidence for bribery was worthless.

Miller's successor was Subrayer who was a dishonest ex-Mysore State Public Works Officer who had been rejected by Morris when he appointed Miller to the position; he told Morris that Miller had left no building records when in fact Miller's records were under a heap of garbage in Subrayer's office. Morris would fire Subrayer barely a year later. The buildings committee had also acted strangely. It had cancelled the building contract with Skipp but there was no reason given for this dismissal. Their action was based only on the statement of the auditor that the liabilities exceeded the estimates by an amount that was so large that it needed to be carefully examined. It was clear that none of those individuals involved in these actions against Miller and Skipp had any idea that they had acted illegally and that the Institute might have to pay very heavily for their actions.

Now, Morris wondered how the auditor's report was accepted without any criticism or even checking of its so-called facts. Rudolf said he was ill while the auditor was in Bangalore and that the resolutions had not been drawn up in a council meeting.

How then were resolutions drawn up? Apparently, they had been drawn up by the Resident Sir Hugh Daly and sent to the Institute; Morris found the originals in Daly's handwriting in his office!

It seemed also that Ghandy, being an attorney, realized that what was contained in these resolutions was libelous and that statements in the auditor's report were obviously incorrect. He wrote to the Officiating Director about them and also tendered his resignation from the council, as seen in part in the following comments:

> " . . . your invitations to members of the Council to make such alterations as they think proper in the draft minutes seems to me to be futile with regard to the Resolutions, as they are already sent to Dr. Travers and to the Government of India respectively.

> With regard to the Resolution III may I enquire whether the wording of it is the same in which it was drafted at the meeting? I am in accord with the Resolution so far as it asks for an explanation, but I am not sure whether

it is right for the Council in a Resolution inviting explanation to prejudge the question. For instance instead of the words 'irregularities which have been brought to light' I would say 'irregularities alleged in the Auditor's Report' . . ."

The document that had been sent to Morris in London ended by saying the actions of the Institute Director, without referring to the council, were, at first sight, evidence of maladministration. Initially, the council thought that the affairs of the Institute should be the subject of an enquiry by a special committee appointed by the Patron; however since this was exactly what Morris wanted members of the council soon spared no effort to prevent an enquiry from being held.

Later, Morris met with Rudolf and Hay and acted as if nothing had happened. It was now his policy to re-establish peace and order and to leave witch-hunting to the special committee when and if it would be formed. However, it was very distressing to all those involved with the Institute that building work on the Institute had stopped.

Student grievances

Now, a new difficulty arose. On November 6th, Morris was asked to receive a delegation from some students in the Department of Applied Chemistry, who wished to present a complaint. Morris said clearly that he wanted to meet with the students individually and not in a group. Though he noted he was at their service, not one came to see him. Morris then waited for two days and not hearing from the students sent for the two who had first come to him and asked them if they had put their complaint in writing.

Since they had not written down their grievances, Morris spoke with them and paid careful attention to what they said. The overall situation with these students was rather delicate since not one of the students who had applied for admission to Rudolf's department was initially accepted. Since there was a demand in India for men who could carry out routine technical analysis, act as laboratory assistants or handle the simple tasks seen in a manufacturing plant, Rudolf had suggested, as an experiment, that he use the facilities available in his laboratory to give such instruction and practice. The senate did not like this idea; but ultimately it was agreed that the men be admitted as apprentices and not students. But, Institute staff made the mistake of not separating them entirely from the other students; this was made even more difficult since they had B.A. degrees.

Morris looked through the records of the two students who had first come to him. D. was the son of an ex-Minister of Baroda State who had a reputation for intrigue. There had been doubts about him being admitted to the Institute but Bhabha had personally vouched for him. A. was the best of the group but his fellows regarded him as 'a little peculiar'. Morris interviewed the two men and got nothing out of them other than to see A. collapse in tears.

Morris then summoned Rudolf who had no idea that there might be unrest among his students. Rudolf thought both these young men lazy and incompetent; they would never use their hands for technical work and looked on while the laboratory workmen and coolies carried on plant operations. Following this, Morris again spoke with the two students and saw more tears. But he found out that members of the council were *also* interviewing them and sympathized with them and with their chief grievance that they were not being "taught industries".

At this stage, as arranged with Rudolf, Morris made several surprise visits to their laboratory. Typically, he saw the following: a student, in spotless white clothes looking on while a laboratory man was carrying out an operation with a jacked tipping pan and about to remove its contents. Seeing Morris, the student stepped up to the plant, seized the wheel of the tipping gear and strained at it. The pan was fully tipped though!

Seeing this sloth, Morris addressed the students generally and told them they had failed to justify any vestige of their complaints and that they must return to their work where he would watch them *very closely* from now on. Next morning they refused to come to work and they were suspended; they then left the Institute. Interestingly enough, one Muslim student, who had taken no part in the mutiny, came to Morris and expressed his complete satisfaction with what he was doing. This young man also told Morris that a young relative of his wished to enter the Institute.

Morris recorded what he had learned for the benefit of the council but on the morning of the meeting he was blindsided with a 90 page printed pamphlet edited by someone in Bombay. Some of its contents included:

> *Mr. E's grievance:* "The professor called me into his room and said that he wanted me to make pure table salt, and chemically pure salt from the commercial salt. I did everything in this connection, except the determination of the chlorides, which required silver nitrate and pure sodium chloride. Neither was available. The former had mysteriously disappeared from the store, and till the session was over no fresh stock could be obtained. I also made experiments for obtaining pure sodium chloride. I wanted, however, to compare my product with samples prepared by manufacturing firms. No pure sodium chloride was obtained."

Interestingly enough, the bottle of silver nitrate had been stolen.

Unfortunately, the Institute experienced a great deal of simple petty theft from the store of goods that could be easily sold in Bangalore. E. was told that he could make enough silver nitrate for all of his work by dissolving an eight-anna piece in nitric acid. Preparing pure sodium chloride was an easy operation that he could have carried out on his own.

> *Mr. B's grievance.* He had short-circuited the battery used in connection with a calorimeter, and was told to make one. This was an excellent exercise; but he had failed to make a battery. Then "he was told to estimate chlorine, but could not do so as the bottle of silver nitrate had been stolen from the store." His attention was called to the "eight-anna piece trick."

> *Mr. F's grievance.* He repeated the silver nitrate story. "When engaged in trying to carry out the preparation of citric acid from lime juice by a method described in literature" the Director visited the laboratory, "and thought it fit to say that Mr. F. was only fit for a secondary school."

Personal matters

Upon his return to India, Morris made a point of dropping in at the club for talk, a drink and dinner on occasion. When he was alone, he dined there often. Friends told him there that all sorts of rumors had been bandied around about him. The fact that Morris was smiling and that Dorothy and their daughter were soon to arrive quieted down these musings. It also helped Morris that Sir Hugh Daly was very unpopular in the station; Morris acquired significant sympathy from simply being on Daly's "list".

Dorothy, leaving for India a fortnight after Morris did, traveled by Bombay where a friend of Morris's met her. Maye Gray [Dorothy's mother], the nanny and two-year old Dorothy followed and met Dorothy in Madras. The first task of the Travers family was to move in to their almost finished house on the Institute site at Hebbal; the garden though was still basically jungle. To care for the garden, Morris hired a very capable individual who had wide scientific knowledge of many types of plants.

Soon they were settled in and began to entertain though their house warming was delayed to Christmas Day when they entertained 20. After dinner they danced, mainly to gramophone music, using the dining and drawing room floors as their dance floor.

At the New Year they drove over to Sivasamuaram and spent several days there. They also paid Sunday visits to places of local interest such as Nundy Droog, the hill that stood twenty miles to the north. In driving on the awful roads though, tires were a big problem especially with what lay on the roads from the shoes of bullocks. Ideally a single tire rarely had a life of 1,000 miles and usually much, much less.

The Enquiry and the Council

Morris had made up his mind that he wanted and welcomed an enquiry, not only to refute the charges made by the council against him but also to answer the questions:

> What was the object with which the Institute was founded?
> What should be its future policy?
> What were the functions and duties of the Council?
> What were the functions and duties of the Director?

Morris felt sufficiently sure of his position not to be incredibly worried and he lost no sleep over Institute matters.

The memorandum that Morris prepared, dated October 15th, 1912 was sent by him marked confidential to each member of the council. One member acknowledged it and wrote back saying:

> "I note that you mark the paper which you have sent me 'confidential',
> and I venture to enquire what meaning you attach to the expression. Do
> you propose to restrict my discretion in dealing with the affairs of the
> Institute? If you do so, I would ask on what authority you do so?"

Morris replied back by pointing out that the council exercised certain powers but individual members exercised none at all. His object in marking the paper 'confidential' was a caution. The question of libel was paramount since neither the council nor its members were exempt from British Common Law.

The next move was by the Resident asking Morris when a council meeting could be held to receive his explanation of the points made in the resolutions. Morris said that since both the council and he had petitioned the Patron to hold an enquiry, the matter was under judgment and could not be discussed. Daly then worked to arrange a meeting of the council anyway. Morris complied with this request and summoned a meeting but also wrote to Simla to state the facts of what was happening. Some members of the council had loose lips and one of them said that if Morris was obstinate he should be removed from office. Morris was fine with this; if this happened he would then release all the papers about the Institute and a major scandal would be the result that would, of course, be *very bad* for British rule in India. Fortunately, Morris soon received a telegram from Simla:

> "Patron prepared without submission of explanation to appoint Committee
> of enquiry into affairs of the institute."

The council meeting of January 17th, 1913 began with Morris discussing the trouble with Rudolf's students. Morris pointed out that since only one or two candidates for admission to the Applied Chemistry Department were really qualified to enter as students, Rudolf had asked his permission to admit some young men who knew a little chemistry and train them as laboratory assistants. Since there was a demand for this type of man, Morris rather reluctantly agreed to Rudolf's proposal, though two members of the senate were opposed to it. While speaking about it, Morris was rudely interrupted by the D.P.I. who remarked—"I have never heard anything so disgraceful." The chairman then noted that Simla needed to be informed of this immediately be telegram. The telegram was sent and interestingly enough the special committee agreed with Morris's actions and praised him for how he had dealt with that situation.

The Resident then asked Morris if he had any explanations regarding the council's resolutions. Morris read the telegram he had received from Simla and Daly again asked him again if he wished to reply to which Morris again read the telegram from Simla. Daly then asked Morris to leave the council meeting to which Morris replied:

> "Under our Constitution, the Director is a member of the Council and Recorder. If I leave, this becomes a meeting of private individuals, of which there can be no official record. The room is at the disposal of those who wish to stop."

Rudolf then said that this was making the matter rather personal to which Daly said that it was so. Daly was shaking with anger and showing here a *terrible* judicial spirit.

Morris left anyway and was later told by Rudolf, who remained, what happened. A resolution, already drafted, was presented by Daly charging Morris with: the responsibility for the chaotic state of affairs (created mainly by the buildings committee); having admitted undesirable students to the Institute; mishandling the student strike and showing that Morris was incompetent and unfit to hold the office of Director from which he should be removed without employing the machinery of a committee of enquiry.

Though the council did not suspend Morris here, Daly made another effort to get rid of Morris without the enquiry being held. In the Institute's government there was a standing committee of the court, which could petition the Patron to exercise his authority on any urgent Institute matter. Daly was the chairman of this body and was also a member of the Madras government at the moment. Daly then went to see the Patron.

CHAPTER XVII

The Special Committee and Family Life

Council Meeting—March 1913

Morris did not believe that Daly's activities in Madras would have any influence on the course of affairs and he turned his attention to laboratory and internal matters. In February the contractor Skipp filed a lawsuit in the Resident's Court in Bangalore. Morris was notified of this since he was the Director of the Institute and he duly circulated this information to the members of the council and suggested that a meeting was needed to deal with this as well as some other Institute matters. This meeting was set for March 18[th,] 1913.

Discussion of the minutes of the last meeting [January 17th, 1913] occupied the entire morning. The next item talked about at the meeting was an idea of Hay's who wanted to add courses in mathematics and applied mechanics to deal with the deficiencies his students had from previous instruction. The D.I. now intervened; he differed with Hay on what was needed. Morris, in a very diplomatic way, said that Hay was the expert on this matter and that members of the council had many other general and administrative problems to deal with rather than to intrude into the details of the work of the Departments of Applied Chemistry and Electrical Engineering.

It was critical for the Institute that its two governing bodies—the council and the senate—keep to their respective duties. Interference by the council in purely academic matters not only was dangerous to proper governance; it was a colossal waste of the council's time when it needed to deal with more important matters.

Finally, the council got to the lawsuit Skipp had brought against the Institute. Morris opened discussion by reading a very short note stating the bare facts of the matter. The Resident then read a prepared resolution stating:

> "That the Director be authorised to defend or to institute any suit or proceedings for or on behalf of the Council. In this connection he should seek advice from the Buildings Committee."

No member of the council had anything to say about this resolution; but it was fascinating that Daly had invited an Indian lawyer to attend the meeting.

It was obvious that no council member, except for Morris, had any idea how serious the situation was. The Resident was Chief Magistrate in the Civil and Military Station, and normally would oversee matters when the case came before his court. In this case though he was the star witness for the defense!

This whole situation was tragicomic; but Morris knew that Skipp would win his lawsuit and serious damage would be done to British officialdom in India as a result. Morris, as an Englishman, resolved to do all that he could to keep Skipp's case out of court. For this to occur though, the government would have to step in.

The Body of Enquiry

Near the end of March Morris received a document from the Education Department that read:

> I, Charles, Baron Hardinge of Penhurst, Viceroy of India, Patron of the Indian Institute of Science, Bangalore, having received from the Standing Committee of the Court of Visitors of the Institute a Report to the effect that they are of opinion that certain matters in connection with the Institute require investigation and enquiry, do hereby appoint a special committee in pursuance of Regulation 35 of the scheme for administration of the properties and funds of the said Institute, to make investigation and enquiry upon the following matters, namely:

> 1. The investigation of the Institute since it was created, and whether any, if so what changes are necessary in the constitution of the various governing bodies in order to secure efficient administration and to avoid conflict of authority.
> 2. The finances of the Institute, including the need for more definite rules regarding expenditure.
> 3. The complaints of certain students in the Department of Applied Chemistry and of certain students of the Department of Organic Chemistry.
> 4. The suitability of the arrangements that have been made for the framing of syllabuses and timetables.
> 5. The conduct of Dr. Travers as Director, and I direct the Special Committee be constituted as follows:

President:

    The Hon'ble Sir Henry Daly Griffin, Kt., B.A., Puisne Judge, High Court, North Western Provinces.

Members:

    The Hon'ble Sir Francis J.E. Spring, K.C.I.E., Chairman and Chief Engineer, Port Trust, Madras.

    Dr. Gilbert T. Walker, C.S.I., M.A., Dsc., F.R.S. Director of Observatories.

    Mr. V.C. French, Deputy Auditor General, Works Department.

The secretary to this body was Morris's former secretary, Sundaram Iyer, whose dismissal from working with Morris was still a big mystery. Morris liked Iyer and thought that he was "attached" to him.

In acknowledging the Viceroy's document, Morris asked that the proceedings of the enquiry be made public as had occurred with other enquiries held under similar conditions in North India. This request was refused. However, the appointment of a High Court Judge as chairman gave promise that the enquiry would be judicial and fair. Morris did not feel trepidation at testifying before such a body though he was concerned about the training and experience of the chair, Sir Henry Daly Griffin, who had held nothing higher than magisterial posts before reaching the bench.

Special Committee Hearings

The committee assembled in Bangalore and first met on April 17th; with the Resident being in Ootacamund. Morris had instructed Professor Watson to call on Sir Griffin and act as a channel of communication for him. Morris was asked to meet the committee on this day and he, as always, kept notes of all that happened.

After formal introductions were made, Sir Spring stated that the Public Works Department should have handled the building work etc. Morris replied by noting that the papers he had received before leaving England said that the building work would be done by the Mysore State Public Works Department. When Morris arrived in Bangalore and met the State Chief Engineer, he was told that he had heard nothing about this and his department was overloaded with work.

It was thus necessary that the Institute have its own architect and building firm. Stevens & Company had been hired as the architects and Morris himself, under the authority of the 1908 provisional committee, created the building entity. This was necessary even though it meant more work and responsibility for Morris. Sir Griffin said that the special committee understood what this had meant. Then, Sir Spring said that the dismissal of Miller without requiring him to turn over his office and papers to his successor had been very ill advised. Morris noted that

> "It had been done while I was away in England, and I did not even know the reasons for it."

Morris here thought that the committee believed he had taken over the building work on his own as well as being responsible for the fiasco of the firing of Miller. The committee here asked that Morris give them a list of names, addresses and occupations of everyone, including students, who had been connected with the Institute since it began.

April 19th had the members of the committee take a tour of the Institute's buildings in the early morning. After meeting Morris later that morning, they asked him for additional papers of interest. Over the next three days, the committee talked to six students but did not think it worthwhile for Morris to attend. On April 22nd, Rudolf was interviewed and told Morris afterwards that he felt quite satisfied that the committee took their side in the matter of the trouble with the students.

Sir Griffin then sent Morris a message through Rudolf, asking that Morris withdraw his "attack" on the Resident. Though he disagreed with this idea, he complied and wrote to Sir Griffin withdrawing all personal remarks. Morris was told the following morning by Sir Spring that his letter had been accepted.

Institute staff [Watson, Tacchella, Kann, Matthews and Rudolf] then appeared before the committee. Their interviews dealt with courses for the students; Morris could have told the committee that small numbers of students and varied qualifications made it necessary to arrange individual rather than group instruction. Kann was even asked what he thought of Morris as Director.

Rudolf seemed quite satisfied with what had occurred in his two appearances before the committee. When the committee told him that they did not wish to see him again, Rudolf then made a very poor decision and left India and began his leave in England. It was mystifying to Morris how the committee allowed the potentially most important witness to leave at such an early stage of its work.

While the committee remained in Bangalore, Morris, seeing very little of them, was on occasion asked additional questions or provided requested papers. The D.I., past and present Dewan were also questioned; the current Dewan was negative about Morris while the past Dewan spoke very highly of Morris. In early May, the committee, having moved to Ootacamund questioned the Resident, the D.P.I. and again the D.I.

On May 21st and for the next three mornings, Morris was before the committee at Ootacamund. He noted that:

> "I have never been in my life in a more difficult position. I knew that one of the matters to be enquired into was 'the conduct' of Dr. Travers as Director of the Institute."

A large number of witnesses had been questioned but not one had been cross-examined. Morris had also been assured that no members of the council held negative views of him; he knew this was an outright lie.

Morris was asked if he wished to make a statement. He felt that it would be best if he gave an account of the history of the Institute from the beginning, and of his connection with it, which he proceeded to do for about ninety minutes on each of the three days he appeared.

When he began to speak about the events of July/August 1912, the chairman stopped Morris and asked why he refused to answer the questions addressed to him in the council's resolutions of August 1912. Morris replied by saying that the council had submitted copies of their resolutions to the government of India, specifically to the Patron. Morris had done the same with his memorandum of October 15th, 1912; this memorandum contained an answer to all the charges made by the council. Since both parties had approached the Patron, these matters were under judgment. Morris, good-naturedly, offered to answer any questions on these matters that the committee might ask. The chairman then closed the session and then said to Morris:

> "I am afraid we have given you a rather bad time. However, we have endeavoured to get to the bottom of the matter, and were obliged to examine you in a searching manner."

This though had not been done. Morris thanked the chairman for his courtesy during the enquiry. While going out of the room, Iyer stopped and whispered to Morris:—"I'm so glad, Doctor, the enquiry is a complete washout as far as you are concerned." Later on that same day, special committee member Walker called on Morris and said—"Sir Henry Griffin has decided that he will not ask you any more questions on the subject of your memorandum of October 15th."

With nothing further to keep them at Ootacamund, Morris and the committee returned to Bangalore at the end of May. While in Bangalore, the committee met with Sir Dorabji, Bhabha, Padshah and Bilimoria.

Morris, on a morning on mid-June 1913, took the committee around the Institute's buildings. Little was said though Morris did note that it was a pity that they could not visit the Institute in July when students were at work.

Two days later, Morris received a note asking that he call on Sir Griffin who told him—"I think that we should tell you what we are going to report to H.E. [His Excellency] the Patron." Griffin then read the following from a note:

> 1. "We think that in the first place you should have acted in accordance with the Resolution of the Provisional Committee (1908) of November 12th 1908, and circulated a Report in accordance with the Resolution passed by the Provisional Committee (1908).

2. We consider that you should have summoned a meeting of the Committee early in 1909.

3. We consider that you must try and cultivate better relations with the Tatas; you must do so if you are to stay in India."

Morris immediately asked who gave the information on what had happened at the meeting of the provisional committee. The name Bhabha cane up as the source. Morris said that this had occurred nearly five years ago and that all members of the special committee had a copy of the minutes of the 1908 provisional committee meeting. Morris recalled that at this meeting Padshah complained that Morris had refused him information; a claim that was immediately refuted by the committee's chair Stuart Fraser. Morris felt certain that, if he had been asked for a report, he would have written one with the help of the minutes and council papers. The idea of Morris looking up this matter and then writing Sir Griffin was discouraged.

Morris never even received a hint from government that his action in carrying on from November 1908 was not approved of. In March 1910, he had taken the responsibility not only of carrying on when the council had been 'sterilized' but also in finding the money to do so.

As to calling a meeting of the 1908 provisional committee in early 1909, Morris had talked to the Resident [Stuart Fraser] and Dewan [Madahava Rao] about this request. Fraser was initially in favor of holding such a meeting but ultimately agreed with Morris's view that such a meeting would only have extended the quarrel with Padshah. The Dewan also said

> "Padshah refuses to tell us what is the alternative to our scheme, so what is there to discuss?"

For the third point of the special committee's note, Morris could say that he had deliberately toned down his accounts of personal differences between himself and the Tata brothers. He had not, and would not, note how the Tatas and Padshah had, when they met in London, deliberately deceived him as to their real intentions.

Personal Matters in May and later 1913

While they were in Ootacamund during May 1913, Morris and Dorothy lived quietly since she was pregnant and due in September. No one in the town seemed to know of the existence of the special committee of enquiry and Morris never heard it mentioned at the club. Since they had Morris's car with them and had excellent roads around them, Dorothy, Maye Gray [Dorothy's mother] and Morris usually left the station in the evenings and rode in the country. Morris also had some time for fishing for carp and rainbow trout in the river even though it was not fishing season.

With Dorothy expecting, they did very little entertaining and lived very quietly through the remainder of the summer and into September. Dorothy had a rough time at the end of her pregnancy. When the wonderful event came the I.M.S surgeon, though a very able man, was not a gynecologist and Dorothy suffered for quite a while after Robert Morris William Travers was born sound and healthy on September 16th.

Robert received a hearty welcome not only from family but also from Morris's Indian staff. This made Morris feel that his staff was really fond of him. Some of their well wishes were quite quaint; Morris's Eurasian mechanic, representing Morris's laboratory servants, made a speech saying that he hoped that "the little master would grow up to be a better man than his father."

Maye Gray left for home at the end of October and early in November Morris's mother Anne joined them. Anne was quite pleased to be greeted by an orderly and personal servant in Colombo who accompanied her to Madras where Morris met her. Anne had a very good time in her three-month stay with Morris, Dorothy and family. They dined out frequently and gave several dinner parties and had a large party for Anne on her birthday of December 1st.

Musical evenings were started again. Though there was no pianist in the same class as Dorothy, Metcalf, a physicist at Bangalore Central College, and a lady in their station were talented violinists and there were several talented vocalists.

Morris's happiest recollections of these days were of family life in his happy and comfortable home sitting in the garden with the children in the evenings, and of strolls in the countryside. Institute worries did not intrude, for Morris was able to put them off when he got home for a meal or for the evening after giving Dorothy the story of his day.

CHAPTER XVIII

Findings of the Special Committee

Meeting of the Buildings Committee

Early in June 1913 Morris summoned several meetings of the buildings committee. The Resident, touring other areas of the States of Mysore and Corgi, was not able to attend any of these meetings but did receive copies of records though he offered no comments. The D.I. came to Morris's office intermittently but spent most of his time criticizing the Department of Applied Chemistry.

French, representing the special committee, gave a great deal of time to the work of the Institute and mediated differences between Morris and the contractor who was fair, co-operative and gave full access to his records. When the work was just about complete, the engineer to the council accepted Morris's suggestion that he leave the Institute. Morris then searched his office and found Miller's missing records; though the special committee regarded these records as lost, Morris now could later recommend Miller strongly to other potential employers.

When the work was done and the bills came in, the cost of the buildings, equipment and services was only about 10% more than the estimate of 1909-1910. Capable experts in the area of building construction deemed this praiseworthy.

The buildings committee awarded the contractor Rs.100,000 damages for the illegal cancellation of his contract and gave him a fresh contract for the completion of the central building and other projected work. The capital balance was now sufficient to complete the central building. Extension of the Applied Chemistry Laboratory and the construction of new bungalows would have to wait though until more money was available.

Council Meeting

Now, Morris had to call a meeting of the council and put before it the findings of the special committee and the new contract for building. On July 28th, 1913 the council accepted one and approved the other. Morris knew several of the members of the council had hoped that the special committee's report would be negative towards him; but it now seemed that this was not to be so. Morris felt relieved at ending this difficult time and actually had some time to attend to research where two of his best students, R.C. Ray and N.M. Gupta, were making good progress on a new class of compounds of boron, oxygen and hydrogen which Morris had so termed boro-hydrates. Such students as these justified, in Morris's mind, his efforts in science in India.

Before he went home in April 1912, Morris had all the "machinery" of the Institute in working order. Despite all of the turmoil, it had remained in good order. Morris though had to rebuild the building organization and began to do so by hiring a good, willing Indian-born and trained engineer who was almost as good as Miller. The contractor, Skipp, was just as anxious as Morris to get work going and began just where the building work was left off; but without the haste seen in 1910 and 1911.

The next item of importance to deal with was a telegram from the Education Department, Government of India—Simla, dated August 22nd, 1913, which read:

"As a result of the enquiry into the finances of the Indian Institute of Science and of arbitration between the Council and Mr. Skipp, the Committee of Special Enquiry constituted by His Excellency the Patron's order dated the 13th March 1913 suggest that the Govt. of India should allot sufficient money to the Institute in the next two years to allow the completion of the remaining buildings, and to satisfy the demands of the contractor under the award of arbitration. It is understood that the completion of the central building and of laboratory equipment and of the extension of applied chemistry together with erecting of remaining bungalows will cost Rs.750,000 of which Council will be able to find Rs.250,000 from revenues, and approximately Rs.100,00 from sinking fund. The government of India are willing to give Rs.200,000 in forthcoming year to enable Council to complete this arrangement with the contractor, provided Council desires this step. Govt. of India are unable to give so large additional sum unconditionally and they will require binding assurance that the money will be spent only on the objects above indicated. Furthermore Govt. of India requires guarantee that occurrences such as the present one will not recur in future years and that they will not be called upon again to meet large expenditure arising out of the management of this institution. Govt. of India will be glad to know at an early date whether the Council accept these terms, and whether they feel legally competent to do so. It is understood that the money must be given, and that the contract must be renewed with Mr. Skipp by September 30th."

Morris was happy with the first part of the telegram since it was addressed to him alone. It also was a good indication that the special committee report would not be negative toward him. It also expressed approval of the policy of using surplus income for capital expenditure, which had been the cause of some of the conflict with the Tata brothers.

But the second half of the telegram implied that Morris had mis-handled the finances of the Institute. Now that all had been sorted out it was clear that this was not the case; but, the report of the special committee had confused the main issue—Was the alleged large excess of liabilities over assets real or imaginary?

Though he hoped that bygones could be bygones and that all could start fresh again, Morris did not believe Sir Griffin's statement that no member of the council had any ill will toward him. Morris had actually accumulated enough evidence over the past few months that several members of the council regarded him as enemy #1 and Rudolf as enemy #2.

Morris sent copies of this telegram to all members of the council, now including Sir Spring and asked if September 1st or 8th would be convenient for a meeting of the council. Spring and the D.I. replied immediately that either date was fine for them. Then, Morris received *identical* letters from the Resident and from another member of the council stating

> " . . . in my opinion the proposed meeting of the Council should be postponed till the orders of the Patron on the case connected with the Enquiry as a whole has been received."

The remaining members of the council, except for Morris's colleagues at the Institute, followed the lead of the Resident and his associate. Morris knew full well what this meant. However, he went ahead and circulated notices convening a meeting for September 8th, assuming such to be the orders of the Patron. Morris also sent copies of the correspondence to Simla. He soon received a telegram from Simla asking him to come to Simla at an early date with the permission of the council. Morris had a good feeling what this meant.

The council meeting was held as arranged with the Resident, D.I., Spring, Hay, Sudbrough and Morris present. Rudolf was still in Europe and the Tata representatives were not present at the meeting.

At the meeting, Morris was granted permission to go to Simla. A resolution was passed accepting the offer of Rs.400,000. Morris put in a statement that after the settlement with Skipp and payment of all outstanding debts, the Institute still had a surplus income of Rs.220,000. There was discussion on the renewal of the contract though it was agreed upon. Morris felt that he had done right in calling the meeting; the absence of the Tata representatives noted that they had not accepted the government offer, which implied approval of Morris's policy of using surplus income for capital expenditure.

Meeting in Simla

Morris arrived, ill and tired, in Simla on September 13th and made an appointment to see Butler the next day at 4 P.M. On the 14th, Morris was immediately shown into Butler's office; Sharp was there also and stayed through the meeting. Butler began by saying—"I must first of all tell you that the Report of the Committee is against you, in fact they say that you ought to go." To this Morris said nothing and there was then a long pause; Morris did not believe what Butler had just said and that he realized that Morris knew at least a bit about the Report. Then Butler completely changed his tone and said—"You can't get on with the Tatas, nor with the present Council." Morris replied by saying no he could not.

Morris now was made to understand that the Resident had wired Butler that he would resign from the council if Morris continued as Director. Daly was under the Foreign Member of Council and had very influential backing in Simla, particularly with the Viceroy himself. Butler had to consider the question—Was Travers or the Institute worth an inter-ministerial fight?

Butler, who was trying to make the best of a bad situation, was also trying to do what was best for Morris. He said to Morris:

> "You told the Committee of Enquiry that you didn't intend to remain in India for longer than ten years, the first period of your service. Now your appointment dates from November 1906, so In November 1914 you can take two years leave preparatory to retiring, drawing vacation pay, and your pension from November 1916. I propose that you be given six months extra leave, so that you can leave India in March of next year. I also suggest that the Report of the Committee remains confidential to the Patron."

Morris said that if he were to resign and no statement as to why he did so was made, people would assume he had been dismissed. Butler replied saying:—"We will make the matter clear in a press communiqué. When you return to Bangalore, say nothing of this." Morris left this meeting admiring the cleverness of Butler.

'Altering' the Institute and the Resident

On Morris's return to Bangalore, he told only Dorothy of what had occurred and been said. It was though clear that council members knew as well as did the two Madras journals and the Bangalore papers [which published laudatory articles on his work]. His former secretary, Iyer, called on Morris and asked why Morris was leaving the Institute since the whole enquiry had been a waste of time; Morris replied by saying that he had finished his service in India and he would be taking leave before retiring.

Next, came a note from the Resident who had received a letter from Simla that he wished to read parts of to Morris. Morris called on him and both men were *extremely formal* as Daly read the letter to Morris.

It was apparent that with Morris resigning, the whole purpose and construction of the Institute would change. Excerpts from the letter included: the closing of the Department of Electrical Technology and the transfer of its staff to a new Engineering College that was being built in Madras; the dismissal of Rudolf who was not considered suitable for a professorship in the new institution and the development of the Institute as a 'trade school' similar to what had failed three years previously in Madras.

Morris's reply to the horrified Resident was simple: he would withdraw his resignation and act in the best interests of the Institute and that the Resident should speak to Rudolf who Morris knew could take care of himself.

Morris also wrote a letter to the Education Department beginning with his rescinded resignation. Daly, in reading the letter to Morris, had said that the 'trade school' idea would be more in accord with the wishes of J.N. Tata than what had occurred. Morris noted in his letter to the Education Department that the intentions of J.N. Tata were clear; all you had to do to see this was to read his April 8th, 1904 letter and see that he wanted an institution that provided for advanced study and research. Morris continued his letter by noting Padshah's University of Research, the Ramsay Report, the Masson & Clibborn Report and how Sir Charles Martin's advice helped J.N. Tata to accept the Masson & Clibborn Report. The Secretary of State had appointed Morris on the recommendation of the Royal Society to carry out this scheme since Morris had the special qualifications and experience to do so. Morris also said that the Tata brothers and the council had made his work much, much more difficult. Morris now noted that he would appeal his situation to the Secretary of State for India through the Royal Society.

In reply, Morris received a very angry letter from the Education Department stating that Morris might have to leave India immediately. This did not frighten Morris at all; the letter also said that the trade school idea had only been a suggestion, put up for discussion by the council of the Institute. Morris was happy; he knew that he had killed the 'trade school' idea and that his policy and plan for the Institute would succeed whether he was there to carry it out or not.

After Morris met with the Resident, it was Rudolf's turn to do so. With Morris present, Rudolf bluntly started out by saying he was not a civil servant and if the council wanted him gone they could do so, either by mutual agreement or through conditions fixed by the High Court in Madras. Rudolf went on to state the terms on which he would resign and they were: that he should receive £6,000 as compensation for the loss of his appointment and that he also be appointed honorary consulting technologist to the Institute.

Morris seconded Rudolf's demands and said that professional appointments such as Rudolf's were held "for life or until fault" subject to an age limit, of course. If Rudolf's case went to court, the council would have no chance of winning. The Resident was clearly shocked but realized that the Patron could not ignore the High Court and that Rudolf was not bluffing.

The next day Morris had another meeting with the Resident and told him that he would be appealing his situation to the Secretary of State through the Royal Society. Morris cited the following events, all since the Coronation Durbar:

1.   Daly's conduct to Morris since the Durbar;
2.   That he had personally drafted the Resolution of the Council in August 1912;
3.   His insistence on Morris answering the charges made by the Council in spite of the veto by the Patron;
4.   His statement at the Council meeting in January 1913 that "so far as he was concerned the matter between us was personal;"
5.   His attempt to secure my dismissal without an enquiry through the Committee of the Court of Visitors;
6.   His leadership in the opposition to Butler's attempt to secure peace which was the background of the offer of the grant of Rs.400,000 and
7.   Morris made specific references to the conduct of certain members of the Council towards himself, to their loose talk, and to the rumors, which they had created.

Daly listened and said nothing. Then he got up, extended his hand and said he wished he had known all this before; now he could better understand the position. Morris shook his hand but thought Daly had no idea what would happen to him if the whole story were told to the Secretary of State for India.

Morris knew that this situation would never have arisen if Stuart Fraser had still been the Resident. But Daly was an ill man, still suffering from his previous appointment as Governor General in Central India. He had a significant problem in remembering meetings, dates and important matters.

It was obvious that Daly felt he had backed the wrong horse in the race and his approach to Morris became more and more friendly to a degree that sickened Morris. Even though Morris wished to conduct himself as a gentleman always, he found he could not do so with Sir Hugh Daly.

Rudolf and Orders of the Patron

At about this time Rudolf went to Simla to plead his case and had several meetings with the Deputy Secretary in the Education Department. Morris had told Rudolf not to speak of him in these meetings and when Rudolf returned he brought with him a paper signed by the Secretary stating that the Patron approved the payment to Rudolf of £5,000 together with his full retirement allowance and the honorary consultantship on the staff of the Institute that he had sought. However, this deal was contingent on both Morris and Rudolf resigning their Institute positions. Both Morris and Rudolf could find no reason for this; Rudolf thought it was due to the Tata brothers while Morris thought it had to deal with the 'groupthink' that pervaded India after a superior had made a decision on a line of policy.

In early December 1913 Morris received, through the Resident, a copy of a "STRICTLY CONFIDENTIAL" letter from the Secretary to the Education Department that contained the orders of the Patron on the report of the special committee. Some paragraphs were in inverted commas; so it was not the original report, which Morris never saw. How much of the report was directly from the work of the special committee and how much of it was the interpretation of the secretary was a matter of conjecture; the secretary, by the way, knew little, if anything at all, about modern university education. The letter from the secretary noted, in part:

> In July 1912, while Dr. Travers was on leave, it came to the notice of the Council through a statement presented by the officiating Director, that there had been a large excess of expenditure over sanctioned estimates for the buildings of the Institute, and that these buildings had not been completed, the Council thereupon appointed a Buildings Committee, dismissed the engineer in charge, and after receiving the report of the Auditor had held an extraordinary meeting on 26th and 27th August. At this meeting, they passed certain resolutions and reported for the consideration of His Excellency the Patron (copies of which were immediately sent to Dr. Travers in London) asking him to take such action as he might see fit. Meanwhile, the Buildings Committee had found that sum approaching to, or exceeding Rs.700,000 would be required to discharge liabilities and complete the buildings. On the 27th September the Buildings Committee saw no alternative save to terminate the building contract . . . . On his return to India from leave Dr. Travers submitted to His Excellency the patron a series of answers to the Council's Resolutions passed at the meetings on August 26th and 27th, also asking him to take such action as he might think fit . . . .

One would think that the council's resolutions, noting specific charges against Morris [and his reply] would have been the main subject of the enquiry by the special committee and that the first question would have been the statement by the auditor that the liabilities exceeded assets by an excessive amount. Morris, had as noted previously in this work, saw that the auditor had included *future projects* as liabilities. The special committee also had come to this conclusion rapidly and that there was no basis for the passing of the resolutions nor for the subsequent action taken by the council.

But the secretary's letter and the report of the special committee made no further reference at all to any of the matters referred to so far, on which both Morris and the council had petitioned His Excellency the Patron to base his enquiry. Morris knew he had been cleared and thus the council had been in the wrong.

Dealing with the letter of the Secretary

Before dealing with the Morris's responses to the main part of the secretary's letter, it was important to consider certain facts:

i.   Morris had been completely excluded from the enquiry.

ii.   The enquiry had been held in the long vacation and the Special Committee had had no opportunity of seeing the Institute at work.

iii.   They had not had the opportunity of realizing that it was not just another Indian College, but something which was quite unique in India, and represented a great experiment, an attempt to create an organization providing facilities for advanced study and research, for resident students.

iv.   It had been in operation for two sessions only, and was working smoothly and efficiently in spite of minor troubles with a few students.

v.   The Council's action had had no influence on the work at all, except that it had delayed developments, and had made the handling of a few inefficient and hard to manage students difficult.

The part of the secretary's letter, which actually related to the report of the special committee read:

a.   "Administration of the Institute and any changes necessary in its constitution . . . . The proceedings of the Senate should be laid before the Council at its next meeting."

This was not the case though. Each and every member of the senate had the right to have his opinions recorded. The council only needed to know the conclusions reached by the senate as a body; this had always been done.

b.   "Approximate dates should be fixed beforehand so that members may be able to arrange well in advance."

Suggestions such as this were obvious but when the constitution made it possible to elect council members who had to travel between 2-8 nights to get to Bangalore for a meeting, it became very difficult, if not impossible to get members to attend at all.

c.   "The annual report should contain an outline of courses of study pursued by each Department, with a note by each professor as to the progress of students under his charge and research work done during the year."

This was noteworthy because it showed that the special committee regarded the Institute as just another Indian College and thus completed ignored what was happening with modern university education.

d.   "There should be a sub-committee of the Council (selected from the members resident in Bangalore) to deal with important and administrative matters."

Morris had raised this point on more than one occasion. The Tata representatives were against it, as they could not attend. The real difficulty was with a council elected as provided for in the constitution of May 1909; it was not possible to find members who would serve on *any* committees.

e.   "The Assistant Professor of Mechanical Engineering should be placed in charge of repairs, etc."

This was a terrible idea. The Institute, even in its infancy, was a large enough place that it needed a permanent clerk of works. This person and the Assistant Professor of Mechanical Engineering had totally different experience and training. The assistant professor also had to devote a considerable amount of time to research.

f.   "A Registrar should be appointed to take over routine administration."

Routine administrative work had been done quite well when Morris went on leave by his secretary, the superintendent of the students' quarters and the clerk of the works [Miller before he was dismissed]. Calling Morris's secretary the registrar would not have changed anything.

g.   "The members of Council should have power to requisition a meeting at any time, not merely term time."

This was quite unnecessary and would have led to much lost time and provided for many, many arguments.

There were many other suggestions scattered throughout the letter. Most dangerous of all was that in the future the Director and professors should be appointed on probation. It had been difficult and would continue to be difficult, for many reasons, to staff the Institute. This new condition would mean that no self-respecting man of the necessary caliber would apply for, much less become part of, a position at the Institute.

> h. "The system of accounts has been carefully examined by the Special Committee. It appears to be unnecessarily elaborate for the needs of the Institute."

The Institute was an elaborate organization with a number of spending departments such as administration, building, scientific departments and students' quarters. It was also not possible to estimate accurately for a scheme involving building a small town with gas, water, electricity, water-borne sewage [an unusual feature for India] and unusual features such as laboratories. This had only been achieved by the careful and orderly handling of the work by Morris and his staff, including Miller.

> i. "The Special Committee thinks that the encouragement of research in connection with the application of science to the industrial development of the country will *for some time* be of great importance."

Morris's reply to this was simple—This would be so for all time!

> i. "Apart from its educational value and its direct usefulness, it (industrial research) has the advantage over research in pure science of being easier to carry out because the ground has been less worked over by eminent scientists."

Morris felt that these observations were utter nonsense based on ignorance. The only member of the special committee, or witness, who presented evidence before it, and who was entitled to speak with authority was Dr. Gilbert Walker; his research on global climate required extremely advanced mathematics and could hardly be called "easy".

Opinions about Morris and late 1913

Morris now addressed the least important matter of all, that being opinions of he and his work. These included:

i. *Being overburdened.* The Patron was convinced that Dr. Travers had been heavily over-burdened by an exacting task of a nature other than what he was recruited for and one that since the institution actually opened formed an excessive addition to his legitimate duties.

ii. *Complaints of the Students.* The Patron agreed with the Special Committee, which thought Dr. Travers's action during the strike a proper exercise of discipline that must be maintained in an institution of this caliber. If the Council had not intervened, and thus supported the students, disciplinary measures would not have been needed. The experiment of "admitting" the poorly qualified students would have been ended at the end of that term.

iii. *The conduct of Dr. Travers as Director.* The Special Committee Report noted the work that Dr. Travers had performed under very difficult circumstances. It also praised him for the way he threw himself into the duties of the Institute and how he devoted himself to its development. However, there was a friction between Dr. Travers and others who were also vitally interested in the welfare of the Institute.

Morris presumed that the "others who were vitally interested in thewelfare of the Institute" to be Dorabji Tata, or rather Padshah who was his representative and Ratan Tata who though took little interest in the Institute.

At Christmas of 1913 everything was again working smoothly and Morris was able to put all worries from his mind. His health was good though he had some nasal discharge problems, which made it necessary for him to always carry a couple of handkerchiefs with him. The Civil Surgeon and the Residency Surgeon, both members of the I.M.S., said that Morris should go to England for treatment by a specialist. This trouble, not affecting Morris's general health, was not going to make him leave India if there was any way he could prevent it.

Chapter XIX

Pushed Out the Door!

Christmas and Early 1914

Dorothy, Anne, the children and Morris spent a quiet Christmas at home. They had no Christmas party but dined on Christmas day with Captain Carter, another friend from the 26th Light Cavalry and their wives. Unfortunately, both of these men were to die in combat before Christmas day 1914.

In late January Morris, seen from this time, went hunting for a week with friends and returned well and eager to continue fighting for his position and the Institute. After returning, he was busy in his laboratory with work on the boro-hydrates and other minor projects.

The general management of the Institute was going well and building was progressing. Morris even tried out one of the improvements suggested by the special committee, which was that local advice should be sought in the design for new bungalows. However, these designs, made by a junior Madras Port Trust engineer, were unsuitable.

The Travers family had not prepared at all for leaving India. It was widely known in the station that there had been significant friction between Morris and the council and the Resident was very unpopular. There was also great sympathy for Morris and the slanders he had endured in August 1912 for having to deal with "such a queer crowd".

Morris's mother Anne, Dorothy, the two children and the nurse went home for England on March 2nd through Madras. Morris carried on in their absence as usual, generally spending an hour at the club before dinner, sometimes eating there but usually dining at home alone or with a friend brought back from the club.

Education Department and the Resident

Rudolf returned to Bangalore on March 9th, 1914 saying that the Institute's records were not in the Education Department. This was quite curious to Morris who knew that these records had been transferred from the Home Department to the Education Department; he had seen them himself in January 1910. They had likely been destroyed which would have accounted for the nonsense about the "intentions of the Founder" in Sharp's letter of November 24th, 1913.

But an interesting fact emerged from Rudolf's visit to Delhi—someone with significant influence was doing everything they could to force Morris and Rudolf to resign and leave India. Before he went to Delhi, Rudolf had told Morris that a member of the council who Rudolf had met in Bangalore asked him if he knew anything about the Patron's decision to dismiss Morris. Rudolf answered by saying that it probably would not happen to which the council member said—"If Travers does not go some of us will resign off the Council." To this comment, Rudolf said that those members should resign now.

Morris had no doubt that the Resident was still "in" with the official clique and that he was telling government they had to get rid of Morris and Rudolf or face the resignation of members of the council. Official word of this came the day after Rudolf returned. Morris received from the Resident a letter covering a copy of a telegram from the Education Department that was said to be the Patron's decision on the report of the special committee. The critical paragraph of the telegram was:

> The Report of the Special Committee is to the effect that the continuance of Dr. Travers and Mr. Rudolf in the service of the Institute is inimical to its best interests. His Excellency the Patron accepts this opinion, but is advised that the dismissal of these gentlemen, before the full period of their service would not be justified by the report. His Excellency is not prepared in the circumstances to order the dismissal of either Dr. Travers or of Mr. Rudolf. He deems it advisable that every effort should be made to secure their retirement by agreement. Failing this agreement His Excellency is prepared to consider an alternative constitution, under which the business of the Institute could be conducted efficiently, and in accordance with the public interest.

Now, since Morris wanted to resign on his own terms, the council now threatened to resign. Morris didn't see how they could do otherwise. One may have thought the facts might have given the special committee the idea that the existing constitution did not provide for the selection of a suitable council; the council, as pointed out by Masson & Clibborn, should not consist of those who had axes to grind. The special committee should have considered what sort of a council was wanted and how to secure its selection.

Further negotiations

Morris found it fascinating to speculate on what the changes for increasing the efficiency of the Institute might be. However, in returning to reality he began to prepare for what was likely coming; he sent a large number of papers back home to Dorothy in England.

He then called on the Resident who had requested a meeting. Rudolf had, of course, brought a letter from Simla stating that the Patron had approved of the award to him of £5,000, a retirement allowance of £1,000 and an honorary appointment in the Institute. The Resident had, soon after receiving a copy of this note about Rudolf, written a letter to the members of council, except for Morris and Rudolf that "the award will keep Rudolf from talking". Morris knew this from Hay's copy of the letter.

In their meeting, the Resident told Morris that there was a proposal to offer him £16,000 compensation on the condition that Morris severed his connection with the Institute. Morris replied by saying that, if he did retire, it would be on the terms set out in his 1906 letter of appointment from the Secretary of State for India. He would receive the money he would have earned over the course of his service of ten years. Morris also would take two years leave and then draw his pension for life. Morris also noted that these terms were not negotiable.

Morris, acting on the wish of the Resident, summoned a council meeting for March 10th, 1914. The Resident had circulated copies of the telegram containing the orders of the Patron, except to Morris and Rudolf of course. The Resident began the meeting by saying that the Patron wished for Morris and Rudolf to retire by mutual agreement. Morris noted that he had read parts of the report of the special committee that were related to himself and Rudolf. Nowhere was it said that either of them was guilty of negligence, incompetence or misbehavior.

Morris had, however, told the special committee he only planned to stay in India for ten years. Since there had been significant disagreement between him and certain unnamed persons, Morris wished to take vacation leave from November 1st, 1914 and draw his pension from December 1st, 1916; the Patron had suggested that he be granted a few months extra leave so that he could retire during the coming vacation. As for Rudolf, the special committee had recommended the reorganization of his Department and suggested that he also retire. Morris also told of the terms that Rudolf had negotiated for himself with the Patron's approval. Morris next said, to the council, that unless the council agreed on satisfactory terms for both he and Rudolf, the positions of Morris and Rudolf would remain as they were previous to the enquiry.

Morris now suggested to Rudolf, who wanted to leave India as soon as possible, that to save time and effort, he should fight for their interests and would ask the Resident if he could arrange with the council to represent them both in discussing terms of their resignations, with the decision arrived at to be binding on both parties. Morris thought lawyers needed to be involved. All parties involved agreed to this idea.

All Morris had to do, as far as Rudolf was concerned, was to settle Rudolf's retirement allowance and some accounts for garden work carried out by Institute labor at his home. Morris's terms were already known though he introduced two new items now as demands. Since he felt that British rule in India would not last very long, he insisted

> " . . . that all payments made to me should be in sterling, in London, with no deductions whatsoever, and should be a charge on the income from the endowment properties of the Institute".

Next, was " . . . that the Report of the Special Committee should be published in full or should not be mentioned or referred to". Morris preferred that the full report be made public. He also would not give up his papers or refrain from making reference to his experiences in India; indeed he felt that he could do more to safeguard the permanence of his work for India by resigning and remaining watchful, but silent, than by staying on at the Institute.

The Resident had gone to Ootacamund because of the hot weather and when the term ended at the Institute Morris also went there and stayed at the club. However, right before leaving Bangalore, Hay sent to Morris a copy of a telegram that he had received from Bombay that had gone to all members of the council. It read:

> "I HAVE SENT THE FOLLOWING TELEGRAM TO RESIDENT AND HOPE YOU WILL KINDLY SUPPORT MY VIEWS WHICH ARE THAT SERVICES OF DR. TRAVERS AND MR. RUDOLF CAN BE TERMINATED BY THE GOVERNOR GENERAL IN COUNCIL AFTER REASONABLE NOTICE. IT IS CONSIDERED THAT A YEARS NOTICE SUFFICIENT SUCH DETERMINATION OF CONTRACT IN OUR OPINION NOT AMOUNTING TO DISMISSAL. IF SPECIAL COUNCIL MEETING COULD BE CONVENED FOR CONSIDERING THE TERMS OFFERED TO DR. TRAVERS AND RUDOLF"
>
> BABSHA

The coded signature meant Bhabha and Padshah without a doubt. Hay did not acknowledge the telegram; Morris observed that these were the people with whom he had to improve relations with according to the findings of the special committee.

With this information, Morris went to see the Resident. Before opening their discussion, Morris noted that the agreement proposed by the Patron for the "departures" of Morris and Rudolf might be held as contrary to the Institute's constitution that could not be changed without the approval of the Tata brothers. They were very rich men and Morris would agree to nothing, which might involve a legal action brought by one or both of them. Morris now demanded positive withdrawal of objections to the council's actions; he would take no steps to secure it. Morris said that lawyers would be needed and that if the case came to trial it was likely that the Patron's orders would be found inappropriate, at best. The Resident was horrified but agreed to Morris's points.

Morris was only in Ootacamund for a week making his position clear to the Resident and to the lawyer. This finished, he drove back to Bangalore. It was the middle of April 1914 and it was evident that he would not be able to get away until the end of June. His nasal trouble was worsening but his health otherwise was good.

Letters, Boro-hydrates and Hay

Morris's research student Ray, who remained a life-long friend, wrote to Morris noting how terrible it was that Morris was leaving India and that, as a result, he would not be returning to Bangalore and the Institute. Morris wrote him back on March 30th saying:

> "I am sorry that I did not say good-bye to you. I had a lot of work to get through, and I was very tired; so you must forgive me. In the midst of a great many worries, I have much enjoyed the laboratory work during the past three years, and I shall always remember my two Bengali friends Ray and Gupta as being associated with many happy moments".

D.D. Kanga, an Indian professor from the Elphinstone College, who had spent some time in Morris's laboratory while on sabbatical leave from his institution, wrote:

> "I have seen many laboratories, but so far as equipment and facilities are concerned, I may say that I am in love with the Institute, and I wish I could stay here all my life. I may say with confidence that our laboratories are out and out the best. If with such facilities, and with the guidance of the best men, students are not able to do much, it is their misfortune, and not the fault of the Institute".

Work continued on the boro-hydrates and there was just enough time to do another set of experiments. It involved the action of concentrated ammonia on the product of magnesium boride with water. The product, which could only be obtained in minute quantity, had a chemical formula of $B_4H_{12}C_6$. This was the only one of this class of compounds that was a solid.

Upon Morris's retirement, he felt that Hay would be appointed Officiating Director [the next Director would be Sir Alfred Gibbs Bourne who was a noted biologist; he would be Director from 1915-1921]. Morris soon began to show Hay what was involved in being Director of the Institute. Hay had done sound work as Professor of Electrical Engineering and he had carried out a variety of Institute tasks given to him by Morris satisfactorily and he was now to learn that being on the finance committee meant much more than just signing checks!

May and June—1914

In the middle of May Morris again drove to Ootacamund and stayed ten days at the club. While there he read through the agreements for he and Rudolf with the Resident and found them satisfactory. Things were very quiet and Morris lunched and dined daily with a small group of the 26th Light Cavalry Regiment, which was stationed at Bangalore.

When he returned to Bangalore a representative of the architects came to talk to him about plans to complete the central building. The architects were curious whom they would have to deal with next; Morris could not enlighten them on this at all.

On May 25th the Resident wrote Morris saying "that the Tata obstruction had collapsed"—as had occurred with all previous Tata obstructions. The government wanted a council meeting to be held on June 17th to settle matters and asked if Morris would so call the meeting. Morris replied that he would do so but the Tata withdrawal of opposition must be definite.

On June 12th, the Resident called Morris and told him that Lord Willingdon [the current Governor of Bombay and later Viceroy of India from 1931-1936] wanted to see the Institute in the morning. Morris showed Lord Willingdon and Lady Willingdon the Institute and Lord Willingdon remarked that he had no idea that anything of the kind existed in India. Morris's response was that there was nothing like it in India and nothing better than it in all of Great Britain.

June 14th had General Wagshawe, commander of the Bangalore Brigade and a friend of Morris and Rudolf, came to the Institute and remarked on the "queer crowd who formed our Council"; and he also noted that the Resident was useless.

On the 16th Morris felt that it was u not necessary for him to be at the council meeting the next morning since at that meeting the council would pass a resolution offering Morris the terms that had been agreed upon. If they would not do so, then Morris and Rudolf would continue in their positions as they had done before the enquiry occurred. Also on this day, Morris handed over charge of the Institute to Hay and thus began his vacation leave. He reminded Hay to send him the minutes of the meeting and any papers relevant to the meeting since he was, after all, still Director of the Institute.

June 17th had Morris spending the day at the club. Early in the morning he received a paper from the Resident, which included a list of the events leading up to the appointment of the special committee and the events subsequent to it. A reply by Morris to Hay, with a letter to the Resident, noted that the paper said much too much.

Hay, on the 19th brought the minutes of the June 17th council meeting to Morris. These minutes were deemed too sensitive to be circulated even in a draft form in Morris's opinion. Morris told Hay to tell the Resident that all references to the special committee should be eliminated or all references made of Morris and Rudolf must be included. Hay and the Resident would have to settle this. Morris said that his next letter on this would come from his lawyer. Soon, Hay brought Morris a copy of the amended minutes. Some members of the council wanted to make minor changes to the agreements; to this Morris said no changes whatsoever.

June 24th was a day for farewells. Morris's junior European staff, which had hunted with Morris and frequently dined with him on Sunday evenings, presented him with a small gold cigarette case, inscribed with their names as a mark of esteem. Morris gave his Office Manager, Gundu Rao, and personal assistant, Busheruddin, watches. Morris spent the day having photographs taken of the office staff, the laboratory staff and the joint buildings staff.

Morris is 3rd from left in front row.

Morris made arrangements for all involved to get a copy of the appropriate picture. He then finished clearing up the house and sent his car on to Madras.

Leaving India

Morris had dinner with the Kanns at the Institute but would not allow anyone to accompany him to the train station where he drove to that evening by carriage.

Upon arriving at the station, he was surprised to see a large number of the Indians he had worked with or employed over his eight years in India. They were there to bid him farewell and he knew from this tear-filled departure that he would not be forgotten.

He arrived alone in Bombay on the morning of June 26th where he had a chat with Stanley Reed at the Times of India offices. Reed told Morris that Sir Dorabji Tata was furious because the Viceroy and Butler both had refused to see him. This made Morris smile.

Morris put himself up at the yacht club and had his last dinner in India while there. He recalled his first dinner in India with Risley who warned him to not engage in politics lest it be his ruin. It was Morris's refusal to play Padshah's political game, which had been the first cause of the trouble he faced.

To Morris's pleasant surprise, his ex-engineer Miller turned up at the yacht club the next morning. He had seen Morris's name on a list of those leaving India; Miller was in good spirits and had a reasonable job near Bombay. Morris was able to tell him that he had been completely cleared of any wrongdoing and if he ever wanted a strong recommendation, Morris would give him one. Miller was happy to hear this.

On leaving India, Morris did not feel at all depressed. He knew that he had exceeded his own significant expectations with respect to the Institute. Further, the assistance that he had given to the Provincial Governments had resulted in bringing modern ideas into the systems of higher scientific education particularly in Bombay, Bengal and in the United Provinces. The Secretary of State for India had adopted the report on technical education resulting from Morris's work for the United Provinces in 1910-1911 and had ordered the Madras Government to adopt it, without Morris's knowledge. Soon the Institute would be seeing a more qualified group of students. In Morris's opinion, his work had been difficult but he had not come to India hoping it would be easy. He did not believe that he was, in any way, responsible for the events, which occurred while he was on leave in 1912 and he firmly believed that he had acted properly since that time. The welfare of the Institute had been, and was still, his primary concern. Had his health permitted he would have stayed in India but he could not. He would not warn able men to not go to India and the Institute; he would keep silent instead.

CHAPTER XX

In Retrospect—Research on the Boro-Hydrates

Setting up a Research Laboratory

It was not until July 1911 that Morris had any time for experimental work and to open what would best be termed a chemical institute. Morris was a Professor of General Chemistry and with him was Assistant Professor H.E. Watson who was a keen and able researcher. Professor N.S. Rudolf, Professor of Applied Chemistry, was equipping his own laboratory. Professor Sudbrough, Organic Chemistry, arrived in October 1911 and found his laboratory ready for him. Morris had also, in the last few years, purchased complete series of a large number of scientific journals and reference books that were temporarily housed in bookcases in an unused wing of the Electrotechnics laboratory. He felt that he had all that a chemistry researcher would want in the way of literature.

He had designed, built and equipped his own laboratories. The stores were well stocked with chemicals and apparatus. There was a well-equipped workshop that the General Chemistry Department shared with the Organic Chemistry Department and, in addition, a liquid air apparatus and a variable voltage electric generator on site. Lastly, the climate of Bangalore permitted hard work in study or laboratory work during the course of the whole year unlike other places in India. Conditions for research were ideal.

However, Morris had a feeling that he had overlooked something when the Institute actually opened. In due time, after Morris had begun research, and the need called for ice he found that he had a problem [no ice making machine]. Ice only came nightly from Madras with the fish supply and there was always sufficient ice to satisfy domestic needs but if this ice was purchased and taken away, there was none left for other purposes [in this case research experiments].

In late 1911 and early 1912 Morris was busy with the Delhi Durbar and needed to be in Calcutta in January of 1912. He thus had to let Watson be in sole charge of the laboratory; which he enjoyed. In the third term though, Morris was fortunate to take two very able students as his research assistants and they were: R.C. Ray, M.Sc., and N.M. Gupta, B.Sc., who were both from the Presidency College in Calcutta.

Chemistry of Boron and Hydrogen

Morris had no intention of returning to low temperature work; but he thought that there could be no better training in experimental work for students than in work involving gas techniques. The boro-hydrates offered a field where there were few researchers and it presented similar difficulties, which Morris had met in the study of the rare gases.

His group began their work by making magnesium boride (Mg_3B_2). They heated a mixture of one part powdered boric acid (H_3BO_3) with two and a quarter parts of magnesium powder in an iron crucible through which hydrogen was passed. This was put into a furnace, which was at red heat and the mixture burned rapidly. The product corresponded closely to the chemical formula MgO and Mg_3B_2. Treatment with cold water produced a gas that burned with a green flame indicating the presence of volatile boron compounds.

Morris noticed that the liquid obtained by treating the crude boride with cold water was golden-brown in color; likely due to the presence of colloidal boron and was not of interest. The solution though was very strongly reducing and bubbled when dilute acid was added to it with the product gas being nearly pure hydrogen. This experiment was worth following up in Morris's opinion.

The solutions evolved hydrogen slowly when cold and rapidly when warmed so a new technique was needed for dealing with them. Evaporation was done in a small distilling flask, into which the stem of a funnel passed through a rubber stopper. The flask was sealed to a receiver, which was connected through a side tube, and a stopcock with a Töpler pump. Exhausting the apparatus while cooling the flask in ice and the receiver in liquid air allowed evaporation to be done quickly and at a low temperature; the gas evolved was collected through the pump.

Now, they had a puzzle. The solution, when evaporated in vacuum, gave a white solid consisting of a water-soluble boron compound and a trace of magnesium oxide. The solid was weighed and transferred to a platinum crucible containing lime (CaO) and ignited. On driving off the water and igniting again there was no change in weight. Several trials gave the same result.

This puzzle went home with Morris to England and was solved when he returned to India. The residue was a boron oxide containing a little magnesia. It was converted into boric acid (H_3BO_3) through evaporation with nitric acid, but was unchanged by ignition with lime (CaO).

A new technique had to be devised for separating the boro-hydrates from boric acid. It was discovered that if the solution containing both compounds was shaken with precipitated magnesia, or magnesia obtained by igniting the nitrate, in an evacuated and sealed glass vessel for several days, magnesium borate was formed and separated as the metaborate, leaving the solution free from boric acid and magnesia.

In a series of experiments aimed mainly at collecting and examining the aqueous extract, 80 grams of crude boride was treated with seven successive quantities of water at room temperature over a period of six weeks. The results led to the conclusion that the reaction between magnesium boride and water was:

$$Mg_3 B_2 + 6H_2 O \rightarrow Mg_3 B_2 (OH)_6 + 3H_2$$

However, later and independently, Ray showed that at -10 °C both dilute acid and water react with the crude boride to give a product with the formula $H_3 B_2$ $(MgOH)_3$. This was an extremely important discovery because this compound was the parent of all the boro-hydrates they later discovered.

Additional investigation showed that they had obtained a compound with chemical formula of $B_2 H_6 O_2$. Further work showed that there were two isomers of this compound that existed in solutions obtained by extracting the crude boride with water. Morris and his colleagues postulated that boron was tetravalent in these compounds. They had now arrived at the end of the Institute's second session. Ray and Gupta were leaving the Institute, as was Morris. Morris was very pleased with his two collaborators and they had convinced him that his work in India had been worthwhile.

After Ray and Gupta

Morris remained in Bangalore till near the end of the summer vacation. He had brought order to the Institute, building construction was proceeding well and his laboratory was open though only two or three students were in it though he had a research chemistry problem to investigate. He had a large amount of the spent boride available and he proposed to treat it with very strong ammonia at room temperature. The solution appeared to contain a very small quantity of a substance that did not lose hydrogen when kept in an ampule or oxidize when this solution was exposed to the air.

Further investigation noted that this boron oxide had the chemical formula B_4 O_5. This compound was almost totally water soluble and oxidized so rapidly to $B_2 O_3$ when exposed to air that Morris could not determine its molecular weight.

Morris wrote up a paper in the names of Ray, Gupta and himself and sent it to the Royal Society. During a Thursday afternoon meeting a bit later, the Secretary of the Society handed Morris back the manuscript saying

"We can't accept your paper. The referees say that they don't believe that the compounds which you describe really exist."

Morris was astonished but knew that the work done, though difficult, was absolutely sound and he replied:

"Fellows of the Society were so slow to accept the existence of the rare gases, that Ramsay and I were considering publishing our last paper privately. This is what I shall be obliged to do now. I'll send you a copy of it."

Morris did this and noted in the article's preface that the Royal Society had rejected the article.

Working alone in a field was something Morris was used to. The work on the rare gases had not been repeated nor had anyone in England repeated Morris's work on liquid hydrogen. Morris was sorry for what had happened to Ray and Gupta but he wrote to Ray telling him that he left the field to him. Morris felt that Ray had acquired the technique and outlook of the Ramsay school and that Ray's contributions to the knowledge of boron compounds was notable. Ray obtained all the compounds he described, as seen in the book Modern Developments in the Chemistry of Boron, as solids and this work contained a discussion on their possible constitution.

CHAPTER XXI

Medical Treatment, the Institute and England at War

Back in England Again

Morris traveled home through Brindisi in Italy and arrived at Dover on the morning of July 19th, 1914. It was a perfect summer morning and Morris was elated to see Dorothy waiting for him on the pier. He was very happy to be with his wife again and away from India.

He did not feel depressed in any way and was comforted by the sympathy his friends had shown him throughout his ordeal. Morris was a bit sore physically though he felt that Dorothy had suffered much more than he. He desired to put the whole Indian "experience" behind him and start over again; he was after all only 42 years old. Financially, his family was not in bad shape at all; he would be drawing £900 a year until 1916 and then £500 a year for life from his pension. The idea of putting his case before the India Office, to Parliament or through the Royal Society was put out of his mind by the approaching cloud of war and his desire to keep his India experiences private.

Morris also now came under the care of the famous ear, nose and throat specialist Dr. Richard Lake to help him with his nasal trouble. Morris spent a week in the hospital after surgery and it was obvious that a return to India would have been impossible; in fact Morris was to suffer from nasal problems until much later in his life. He was though fortunate that it never stopped him from working.

On July 29th, Morris had lunch with Rudolf and they both felt that war was not coming and any talk of it was foolish. Morris and Dorothy went away for the weekend on Friday, July 31st to Beaulieu while the children went with their nanny to Bognor. The following day unfortunately gave everyone a note that war was imminent. On that Sunday, Morris and Dorothy, while walking to buy the Sunday newspapers, saw large numbers of Territorial soldiers leaving their camps and heading by train to London.

Morris and Dorothy went back to London on August 3rd; the next day Morris went to the India Office and volunteered himself as a scientific man for service at home or abroad for a year and a half in an unpaid capacity. He was asked to come back in a week and was then told that he likely would not be wanted since the scientific staffs of the government were filled.

Correspondence and India

On June 28th, 1914 there was a two-column story in the Madras Times about Morris's departure from the Institute and India. It dealt with the difficulties Morris had faced and the inability of the council to give him adequate assistance. It also spoke of the quality work that he had done and noted that he was leaving without a word of official thanks, but added:

> "The quality of Dr. Travers's work is not to be judged by the amount of mud that has been thrown; the one has nothing to do with the other, and we know the source of the mud, and why it has been thrown."

Soon after he had returned home, Morris also saw a note in The Pioneer referring to his and Rudolf's resignations as "ending an unsavory episode". One could certainly read into what this meant; Morris ignored it entirely.

In addition, a day or two after Morris left India, Hay, as the Officiating Director, wrote to Morris saying that council members wished to add a condition to Morris's agreement that they were in no way personally liable. Morris, amused, would not agree to any revisions. He had no desire to keep up correspondence with India or of interfering in any way with the Institute's affairs. His former staff did of course know that Morris would help them out in any way if he could.

Interestingly, Watson wrote him on November 29th and told Morris about the council's attempt to hire a new Director. As before, the council of the Royal Society appointed a committee to make a nomination for this position. Though the Tata brothers were bent on a poorly qualified Australian, the Royal Society nominated Robert H. Pickard who happened to be someone Morris knew, liked and respected. Pickard was then 40 years old and the Principal of the Blackburn Technical College, which he had just significantly re-organized. He was thus the exact type of man the Institute needed but he was rejected.

Pickard wrote personally to Morris telling him what had happened. After being nominated by the Royal Society and his nomination being accepted by the Patron, he thought he had gained the position. Pickard was later amazed to discover that his appointment and control of the Institute was in the hands of the council; which Padshah was now again a member of. Pickard was glad to be rid of this matter. With this news, the Royal Society refused to further involve itself with the selection of a new Director.

Since no senior man would apply for this post, given retirement age restrictions, a special by-law was enacted to allow the appointment of a director for a short period of time. Two applications were tendered for the position; the D.I. and the D.P.I. The D.P.I., Alfred Gibbs Bourne, aged 55, was appointed. Bourne had spent 20 years in the Indian Education Service and had been a member of the Royal Commission on the Indian Universities of 1902 and was strongly against the D.I.'s schemes for developing industries. Morris felt that Bourne believed he had found an easy job, and having no ideas of his own on modern university education, would make no changes. Morris was right since Bourne's first move was to fill the fifth chair in the Institute with a Professor of Applied Bacteriology position, which Morris had of course proposed to do. Bourne's salary was larger than what Morris had received but the position was not pensionable.

Morris also saw very little of Rudolf. The idea of being partners as consultants was proposed by Rudolf, but nothing came of this idea. Rudolf was to visit Morris in his glass works in 1916 and congratulated him on what he saw; Rudolf soon afterwards was engaged to a lady he had met in Bangalore but, before the wedding, he suddenly died from pneumonia in 1917. Tacchela did not stay long in India though he was, in later years, to lunch frequently with Morris and Dorothy. Watson became Professor in the Institute in 1916 and became Ramsay Professor of Chemical Engineering in 1934 in University College London. Morris and Watson met several times in the years after 1914 and they were always friendly to each other.

Entering the Glass Industry

In a very short time Morris's whole interest centered on the ever-expanding war. He still had fever germs in his blood and any undue exertion, stress or illness brought about high temperatures. He did though serve, for a short time, as a member of a searchlight squad but he soon took on a chill and had to find a more suitable 'occupation'.

Early in September 1914 Morris was fortunate to meet Douglas Baird, the Chairman of Baird and Tatlock of London, which was a well-known firm of scientific apparatus suppliers. After talking they discovered they needed each other; there was already a terrible shortage of scientific glassware in England and Morris needed meaningful employment that could help the war effort.

Since he now might be able to carry out some useful scientific work, Morris asked Professor Donnan to give him laboratory space at University College. However, the Chemistry Department was moving into new laboratory facilities and he regretfully could not grant Morris's request. Morris then asked Professor Brereton Baker at the Imperial College who gladly gave Morris a room in his laboratory. Baker then told

Morris that he was also working on the manufacture of glass in conjunction with Harry Powell, who was the Technical Director of James Powell & Sons, Ltd., of the Whitefriars Glass Works. Morris went to see Baird and told him that he should talk to Baker. A good deal of correspondence ensued involving Baird, Baker, Morris and Powell but ultimately, on good terms, Powell and Baker went one way while Morris and Baird & Tatlock went another way in the manufacture of glassware.

Morris began his work by first collecting and analyzing samples of soda-lime glass for table working and of resistance glasses for the manufacture of beakers, flasks, etc. This obviously involved many, many accurate analyses of glass. With a good deal of practice Morris was able to complete an analysis of soda-lime glass, including the determination of both the soda and the potash content, in 24 hours. From his analyses, and the reports he was given by Baird & Tatlock's expert glassblower, Morris came to the conclusion that the composition of the best German 'table working' glasses had the chemical formulation of:

Compound	% Composition
SiO_2	68.0
Al_2O_3	4.0
CaO	7.5
MgO	0.5
Na_2O	13.0
K_2O	7.0

Potash, as well as soda, seemed to be an essential component of all of the best glasses. Analyzing resistance glass was a more difficult task; but Morris carried out a large number of analyses and came to the conclusion that a glass similar to the Jena resistance glass, though containing a bit less zinc, would be suitable for what he and Baird & Tatlock needed. It would have the following composition:

Compound	% Composition
SiO_2	66.5
Al_2O_3	6.5
CaO	3.5
ZnO	5.0
B_2O_3	5.0
Na_2O	11.5
K_2O	2.0

The next step was to prepare batches and make glass by melting them. Morris did not believe that experiments in small laboratory furnaces would be valuable; so early in 1915 Baird had one of Powell's glassblowers set up a small skittle-pot oil-fired furnace in a long shed which had been built next to his factory in Walthamstow.

Morris then calculated the quantities of sand, alumina, soda ash, etc. for batches corresponding to his above analysis. Glass tubing and a small beaker were then made and the glassblower stated the tubing to be excellent for table working and as good as the best German glass. After the first beaker was made, water was put in it and then boiled.

The next question was what to use for raw materials. Given that all manufacturers practiced formula secrecy, Morris had to begin searching for raw materials without any help. As best he could find, feldspathic rock was a typical source of alumina, and sometimes china clay also, but this clay did not incorporate easily into batches. He knew that ground feldspar was imported from Sweden by porcelain manufacturers, so he obtained some of it, found it reasonable, and persuaded Baird to import 50 tons of it. This gave all the needed alumina and some of the potash needed for the table working glass; but to keep the alumina and alkali in balance in his resistance glass he had to use both feldspar and china clay. Morris discovered that the first step in making good glass was to mix the constituents of the batch thoroughly, preferably by grinding.

Concomitantly, a group of chemists was publishing formulae for the composition of batches from which chemical glassware might be made. These formulae were based mainly on information in textbooks and were obtained by representing analytical results in terms of laboratory chemicals. No suggestions were offered as to what raw materials should be used.

At this stage, Morris and his colleagues, who were engaged in the manufacture of scientific glassware, realized that they were attempting to create a new industry. In actuality, they were in no way aware of what they were facing up to.

Family life

Now we arrive at Christmas 1914. Morris, Dorothy and the children had now all settled in at 17 Lexham Gardens. Dorothy's parents and Morris's mother Anne dined with them on Christmas Day with the children appearing at dessert. Since this was Robert's first appearance at a Christmas dinner, he, a wonderful child, fulfilled a Travers family custom of walking down the dinner table.

At this time Dorothy's health was not good, so in January 1915 she and the children went with the nanny to Bognor; Morris paid them occasional weekend visits. Morris spent his days in the laboratory and later it became necessary for him to spend most of his time at Walthamstow. He typically had breakfast at home and bought food at the Liverpool Street Station, usually ham sandwiches and Bath buns. He would eat dinner somewhere on Liverpool Street on his way home.

During the winter Morris and Dorothy spent a long weekend with Francis and his wife in Bristol and met many old friends. Francis's new laboratory was in full working order and the staff had doubled since Morris had left in 1906. McBain, who had succeeded Francis when Francis had replaced Morris, was doing excellent work on adsorption and with soaps. The new university buildings at the top of Park Street, centering on the great new hall, were taking shape. Morris also, while in Bristol, had a chance to talk with Lewis Fry who was now a very old man.

In the summer of 1915 Morris and Dorothy rented a furnished house near Stoke Poges; Morris joined the family on weekends there. In the autumn and thereafter, they were all together again in London.

It is though amusing to look at what Morris and Dorothy thought of their home. The cook had a full-time job, which included washing dishes and preparing meals for the adults and the children. Initially, Morris and Dorothy had a parlor maid and a housemaid who also helped in serving the evening meal though they soon, to save money, had to let go the housemaid. The nanny looked after the two children. They did not even think of a car and a radio was then only a dream; in addition, Morris did not belong to any clubs. Actually, beyond very limited hospitality, the Travers spent nothing on entertainment.

Morris had started work on glass without a thought other than it was important to the war effort. Baird & Tatlock supplied him with chemicals and apparatus and the Imperial College provided him with free laboratory space. During 1915, he was made the Scientific Director of a new company named Duroglass Ltd. and was given shares in consideration of past and future services to the company.

At Easter time 1916, Morris and family spent ten days at the Bridges Inn near Princetown where they well entertained at the inexpensive rate of fifty shillings a week. Morris fished on Dartmoor even while it snowed. In May, Morris and Dorothy began to negotiate an exchange of homes with people living at Chingford; but their correspondence was with a member of the public relations staff of the Admiralty and he was too busy trying to explain what had occurred at the great battle of Jutland to continue their correspondence; thus it was dropped.

In the autumn of 1916 Morris had to face the fact that the cost of living was rising. Morris and Dorothy only had an income of £650 per year and £120 of that had to be paid for life insurance. They were graced with some good fortune though; the lady living next door to them asked Morris if he would transfer his lease to her since she wished to rent three adjacent houses to form a residential club for female university students. Morris discussed and settled all this with her on October 16th, 1916 and Dorothy later arranged for the transfer of the lease and the warehousing of the furniture. In the future, when these students were on vacation, Morris often spent a night or two in his old home as a paying guest.

CHAPTER XXII

Chemical Grenades and a Fractured Skull

Gas Warfare

On April 22nd, 1915 at 5 PM the German Army released chlorine gas from 5,730 cylinders opposite Langemark-Poelkapelle, north of Ypres. This gray-green cloud drifted across positions held by French Colonial troops from Martinque who broke ranks, abandoned their trenches and created an 8000-yard wide gap in the Allied lines. Though the German Army failed to exploit this breakthrough the era of chemical warfare had begun.

Duroglass, Ltd. had just obtained an oil-fired glass furnace when they received an order from the War Office to manufacture glass bulbs about 4 inches long and 2.5 inches in diameter with hemispherical ends and necks 4 inches long. It was Duroglass's opinion that these bulbs were to be used as a form of British retaliatory measure in gas warfare.

Approximately a month later Baird came to Walthamstow and told Morris that it was proposed to fill these bulbs with a liquid whose vapor was asphyxiating and then seal off the necks. These bulbs were to be used as a hand grenade. This suggestion had come from the War Office who felt that Baird's firm could supply the bulbs and Morris, being a chemist, could arrange for the filling of them.

At first Morris refused on moral grounds to be involved in this task; but the War Office noted how important this work was to the war effort and Morris reluctantly agreed to help. There was yet no actual business connection with Baird in the making of glass and Morris was to have no business connections with the poison gas venture. Morris was simply acting on his offer to the government to serve as a scientific man for the first 18 months of the war without any pay to him. Had he known that he was being asked to implement an idea of one or more men from a body called the Trench Warfare Committee (T.W.C.), he would have declined to participate in chemical grenade manufacture.

The development of the manufacture of scientific glass was already occupying all of Morris's time. He had not realized how much he had to learn and how much

British technical practice would need to differ from what it had been to develop this new industry. Now, with the need for the chemical warfare bulbs, Morris's attention was taken away, to a large degree, from this vital scientific glass enterprise.

First, a shed had to be built to house the machinery needed to make the chemical warfare bulbs. This necessitated designing semi-automatic apparatus by which the glass bulbs could be run through fume chambers, where each was filled with a given measured quantity of the liquid that gave off an asphyxiating vapor. Before leaving the chamber, the neck of each bulb had to be heated in a gas blowpipe flame, drawn off and then sealed. Morris designed the apparatus, which was constructed in Baird & Tatlock's instrument factory, so well that unskilled girls could do the work.

However, when Morris learned the identity of the liquid he asked around and discovered that there was only a one-week supply of it in all of England! The T.W.C. had failed to check on this. Morris was then instructed to use a mixture of this chemical and another. He then told the T.W.C. that in the presence of water one of these chemicals practically neutralized the other. This observation by Morris was met with a stern reprimand—"It was his duty to carry out the orders of the Committee, and not to criticize". The first chemical was then removed entirely from the grenade composition.

At this point Morris and his colleagues were instructed to wrap each bulb in crinkled paper, push it into a round tin box and then solder on the lid. A short copper tube, closed at the inner end, went into each tin box. A soldier was to be supplied with short lengths of time fuse, pinched into a detonator, which was to be pushed into the copper tube. He was then to light the fuse with a match from the box and then throw the grenade at the enemy. It was ordered to pack eleven grenades in a wooden box with one tin containing twelve detonators with lengths of time fuse attached.

With this done, Morris resigned from assisting the T.W.C. on this project. An official was sent to ask him why and Morris responded by picking up one of the grenades and then dropping it on the floor. The official then picked up this grenade and also dropped it on the floor; he said why did no one tell him. Morris said that he was prohibited from criticizing the T.W.C. Fortunately, this was the end of chemical hand grenades but not the T.W.C., which survived this incident.

Chemical Shells

The bomb factory, as described by people in Walthamstow, shut down, but not permanently, much to Morris's disgust. Just before Christmas 1915 Baird was asked to begin filling shells with lacrymatory fluids. Though Morris would much have preferred to not be involved with this project, he was convinced by the T.W.C. that his assistance would be less harmful to the glass factory than if someone else would be in charge of this shell project. So, Morris reluctantly agreed to continue as manager of this venture with two young scientific men as assistants.

There were though some difficulties with this project. The ignorance of members of the T.W.C. on the nature of any construction material other than glass tubing was obvious. An example of this involved the setting up of an apparatus for filling the shells, which was a simple and obvious design. Morris proposed to make it of brass tubing and blanks, soldered together, using gunmetal stopcocks. Morris noted this to the T.W.C. and received the reply—"The liquid corrodes all metal but iron and iron must be used with porcelain stop-cocks." Morris persisted with his suggestion to use brass but was told that he could use steel rather than iron if he wished. The apparatus was then made of brass and painted so that it looked like steel.

In October Morris's two assistants received notices and both joined the Army. The T.W.C. had been unable to decide if they should be protected; unfortunately Morris could not replace them. Early in November 1916 a special order came in for shells and Morris had sole charge of the shell and glass factories.

Accident and Medical Care

On November 16th, 1916, the very day that Morris had arranged to transfer the lease of his house to his neighbor, he was sitting in his office next to the filling factory busy with a government inspector. His foreman came to tell him that there was a man in the yard who wanted to see Morris about a possible assistant manager position. Morris, excited that he might finally have some help, ran across the factory even though the lights had been dimmed [because the ladies who worked in the factory were on a tea break]. Unknown to Morris, one of them had backed a shell trolley between two piles of shells and had left the handle of the trolley jutting out across the gangway, about three inches above the floor. Morris did not see this, tripped and fell and had his head strike a stack of shells.

Somehow, he managed to stagger back to his office bleeding from the nose and mouth. He was then placed on a stretcher and the doctor was urgently sent for. Morris felt outside of his own body but did hear a little man in a brown felt hat say—"We'll send him to Tottenham Hospital. That'll give him the best chance".

Fortunately, an ambulance was standing by in Baird & Tatlock's yard waiting to be fitted up as a mobile laboratory; so they put the stretcher in it and took Morris to the hospital. Two or three days elapsed before he regained consciousness and found Dorothy by his bed. She had slept in the corner of his private room and she stayed with him for several more days. At one point, Morris's surgeon came to see him and when Morris insisted on knowing the truth of his condition was told that he had fractured the base of his skull. He would need to lie perfectly still for several days; this was obvious to Morris who had already found out that if he moved his head slightly the room seemed to spin around and around and then disappear. Morris thus obeyed his doctor.

Through his ordeal, Morris felt that the hospital staff was exceptionally kind to both he and Dorothy. Morris stayed there for six weeks and was confined to bed for all but the last three days of his stay. Dorothy stayed with him for three weeks, likely till it seemed that he would recover. Morris was allowed no visitors but after a week's time a delegation from the filling factory was allowed to see him. They filled his room with flowers. Morris had heard, while hospitalized, from Baird that the T.W.C. had taken over the factory without expressing regret at Morris's accident or thanking him for his services

Period of Recovery

When he was leaving the hospital, Dorothy asked if she could get some official help in moving Morris to the country to further his recover. The reply back said that since he was not an official, nothing could be done. With this distressing news in hand, Dorothy went and spoke with the Station Master at Waterloo. He was extremely helpful and had, when Morris arrived, a porter waiting with a wheeled chair to move Morris to a reserved first-class carriage where an officer of the train company was in charge.

Dorothy had rented a house at Beaulieu in the New Forest where the nanny and children went in the middle of December; Morris and Dorothy joined them just before Christmas. While staying there, Morris picked up strength and by the beginning of February 1917 was able to walk around and about for an hour or so. However, Morris had trouble having his eyes focus together for several months and if he were to turn suddenly, he would collapse to the floor. Dorothy read to him; but had to avoid comedy [Morris would get a headache] and the dramatic.

They returned to London on February 11th, 1917 and after staying a week in a hotel, Morris rented a house in the Archway Road area while looking for a permanent home in the area. In order to get to the factories at Walthamstow, Morris had to walk to Highgate Road Station and take the train to Black Horse Lane Station which took him less than an hour.

In looking for a permanent home, Morris and Dorothy were told that there was a place available in Beacon Hill, Priory Gardens. The agent thought this might suit them and it was not a palace as Morris had joked it would be. Morris and Dorothy got the address and, after some initial difficulty in locating it, found a charming little house facing the Highgate Woods across the road. Fifteen minutes later they had secured the transfer of the lease and they moved in at the end of March. Their neighbors were George Simpson, who had been a member of the Scott Polar Expedition, who Morris had known in India, and on the other side the family of a colonel who was in the army in France and came home occasionally on leave.

When he was well enough to return to work, Morris was warmly welcomed back to the glass works. He was though no longer involved with the work in the filling factory and was, in fact, prohibited from entering it. Morris thought this funny but of course told the Manager of the Filling Factory that he and his staff could come and go in the glass works. Morris still, though in an unofficial capacity, was consulted by those in the filling factory who had carried on the work he had been doing though now with a large staff present. A nurse was in charge of their first-aid station; but she of course handled the occasional cut fingers in the glass works. The excellent cafeteria in the filling factory was able to provide lunch for all in both plants.

CHAPTER XXIII

Specific Items and the End of the War

Beakers and Petri Dishes

From his experience, Morris knew when a problem could be solved by scientific analysis, success in solving that problem carries the next researcher to further success. This is not so though when success is a matter of craftsmanship. Morris's workers were craftsmen and they were willing to try variations of traditional methods in dealing with new problems; Morris though instead insisted on trying something entirely novel.

Practically nothing could be found in the available technical literature [German information was obviously not that available]. By chance, Morris and his glass works were fortunate to get some hints from a pamphlet issued by the Jena firm, Schott and Genossen, which showed glass blowers making chemical apparatus. Details could be interpreted from pamphlet's artwork.

When a beaker was "blown", and removed from the mould on the end of the blowing iron, it looked typically like a thin-walled bottle connected with the blowing iron by a thick-walled neck. Morris's craftsman were able to come up with the practice that touching the neck with a piece of cold iron allowed it to crack through permitting the article to be caught on a tray covered with ashes and then placed in the lehr. The next step was to cut or "crack" the article into two pieces. The "beaker" was then put on a fixed plate and a scratch was made with a diamond on its side and it was then moved onto a revolving plate. On there a flame played on it at the exact level of the scratch that ran uniformly around the item. The top was thrown into a waste bin while the lower part, the beaker, was ready for further finishing.

Baird, who frequently visited government departments [such as the War Office], brought back orders from time to time for glass goods that no other firm in England could provide a supply of. At the end of 1915 he brought to Morris an enquiry about petri dishes. They were needed by bacteriologists to grow cultures. Imported stocks were gone, the demand was urgent and vital and Duroglass, Ltd. just had to make them!

To Morris, there seemed to be no real difficulty in blowing the vessels as flat-bottomed beakers. It was easy to get a flat bottom; it was gained by leaving a narrow gap between the bottom and the vertical side of the mold—this gap was so slight that glass did not enter it as air escaped. The forming of the ball required special care. It just was not possible that the vessels could be cracked off like with a beaker; the glass could not be freed near a point at which the direction of the surface changed so completely free from internal strains by annealing that a crack starting as a scratch [made by a diamond] would run uniformly around the vessel.

Given this obstacle, the making of a petri dish started out with 1.5-inch deep vessels. First, they were cracked off at the top as near as possible to the shoulder, leaving a vessel with an irregular depth of perhaps 3 cm. This item was placed on a turntable and a glass-cutting diamond was mechanically brought into contact with the inside, at the right angle to make a cut at the proper distance from the bottom. On turning the table, the cut made by the diamond went all the way around the inside and penetrated the wall of the vessel. The two halves then came apart and all that remained was to smooth the lower part on a grinding wheel.

Morris spent a long time in developing and perfecting this process. The work was costly but produced the vitally needed petri dishes. Government refused though to compensate Duroglass, Ltd., ignoring the time spent while absurdly pointing out that they recovered the glass from the vessels, which were broken to make the petri dishes.

Additional Laboratory Glassware

An item that required a great degree of skill for the glassblower was a large globular flask with a long neck. These flasks, possibly up to five liters in volume, were very much in demand for laboratory work and were blown from resistance glass. Morris, in fact, kept one of these flasks for a while but since this glass type broke easily, the item did not survive very long. Similar flasks, with extra long necks, and totally free from any blemishes and needed for x-ray bulbs were made from soda-lime glass. Duroglass, Ltd. made most of those that were required.

The manufacture of articles made from glass tubing in the blowpipe flame soon became very important. This work was quite outside the expertise of glasshouse workers. Though there were a few men skilled in this area in London, all were fully employed and not available. So Morris hired a number of young women, teaching them on his own, and started them on making test tubes. These young ladies, ranging from age 14 and up, seemed to do the job better the younger they were! These "experts" then began more difficult tasks within the glass works.

This business of making test tubes called for the expenditure of a great deal of ingenuity on the part of Duroglass, Ltd. It also required that the company deal with a critical market and not accept loss of item quality due to conditions of war.

However, this test tube enterprise certainly brought Duroglass much trouble and very little profit.

Graduated Glassware

In the fall of 1917 Duroglass, Ltd. began to manufacture graduated glassware, which was about the time Morris was sufficiently recovered from his fractured skull of November 1916 to return to work. To do this manufacture correctly, Morris hired a young physicist named Frank Harris as an assistant. Duroglass, Ltd. also entered into an agreement with the Head of the Walthamstow High School to employ high school girls in the glassworks for this job.

Before going into manufacture of the needed graduated glassware items, Morris investigated pre-war supplies of these items with Baird. The quantity of high quality glassware manufactured in Great Britain before the war was negligible, except for thermometers. The section of the National Physical Laboratory that tested graduated ware confirmed this. Fortunately, this facility gave Morris and Harris tables giving the corrections for barometric pressure and water temperatures required when fixing the volumes of vessels by weighing them empty and when filled with water. This data saved a great deal of preliminary clerical labor.

Standardizing a 500 ml flask involved taking the clean, empty and dry flask and putting a streak of India ink down the outside of the neck counterpoised on a balance. The barometer would then be read and the air and water temperatures [of water in a tank in the room] would be compared to the tables provided. The weight corresponding to 500 ml would be noted and that quantity of water would be added to the flask.

The flask was then put on a flat iron plate with a pillar carrying an arm supporting a reading telescope and a ruler. The arm was then adjusted till the straight edge touched the ink streak and was level with the bottom of the water meniscus. A line was then drawn through the ink streak with a scriber and the flask was coated with acid resistant wax. The flask was then put in a lathe and a cut through the wax was made around the neck at the exact level of the mark on the ink streak, again using the sighting telescope. Finally, the inscription on the flask was cut through the wax using a pantographic apparatus.

To determine the volume of an item such as a pipette, Morris devised a new method for his time. He took pieces of nickel rod, drilled them, rounded the bottom of the hole and polished the insides. The surface was finally polished at the open end so that when it was filled with mercury and a piece of polished plate glass was slid over it the mercury content was of a definite volume.

To standardize a 10 ml pipette (as an example), Morris would first make an ink streak along the pipette's stem. The pipette, held vertical in a clamp, had a rubber sheet underneath it. Water was then drawn up in the pipette to get close to where the mark should be; the water then ran out of it in the standard time for a pipette of that size and then its tip was instantly pressed down on the rubber sheet with the pipette being clamped back into position. Mercury was then poured into the pipette through a funnel. Water rose up between the mercury and the glass and a water meniscus formed above the mercury meniscus. A mark was then made through the ink streak at the level of the bottom of the meniscus.

Morris felt that for these standardizations difficulties lay in overcoming faults in such processes as the uniform coating of the apparatus with an acid resistant wax and in the manual skill of the person doing the standardization. He spent about 18 months at this job and had he done it any longer he certainly would have devised a method to mechanize the process of standardization of volumetric ware.

End of the Great War

With the conclusion of the war on November 11th, 1918 Morris knew that there would be big changes with Duroglass, Ltd. The following day Morris was told by Baird that all the orders they had on the books from the British government and the American army had been cancelled. There was no suggestion made to protect home-manufactured scientific glassware; it was clear that scientific glass would soon be flooding England as it arrived from the continent.

Duroglass, Ltd. could not hope for consideration from scientific instrument dealers, who had been earning a normal percentage on sale of English-made glassware, while incurring only normal risks and charges. In the future Duroglass Ltd. would suffer since it did not have the moulds or a large number of skilled workers that manufacturers did on the continent. Their range of goods it made was also not very large.

The Board of Trade put forward a suggestion that government should acquire Duroglass Ltd. and use it for development work. Morris was asked to sign on for a while as Scientific Director. Though he appreciated the compliment, Morris did not view this as a sound project and he declined the offer. He also decided to discontinue his association with Duroglass Ltd. and accepted an offer of £1,000 for the stock that he held and thus "retired".

Morris probably lost between £3,000 and £4,000 from his time with Duroglass Ltd. but he was alive and in reasonably good health at the end of the war. He was proud of his contribution to the war effort, had learned a great deal and made many friends. Now, he wondered what came next in his life.

CHAPTER XXIV

Visit to America

End of Glassware Work and Travel

In May 1919 Morris left Duroglass Ltd. and caught whooping cough from his children. He was a sick man during the summer of 1919 and did not completely recover until he spent a month in the country.

Morris was quite perplexed what to do next with his life. He felt that he could not renew his connection with academe and he only had industrial knowledge in the manufacture of scientific glassware. After thinking a while, he decided to extend his industrial knowledge by going to America and spending a couple of months there. Morris felt that his connection with the glass industry and his academic reputation would serve as appropriate introductions for him.

Strangely enough, he found it exceptionally difficult to find information he wanted about America and its glass industry and the Board of Trade could tell him nothing. Morris therefore decided to book a passage for himself to New York City and explore America on his own. It was only by being a pest that he was able to secure himself a passage; he sailed for America on September 8th, 1919.

East Coast Cities and Works

On arriving in New York City, Morris found that a letter from the Board of Trade saved him from paying a duty on his typewriter but this letter was useless for the rest of his American travels. He set out for Washington, D.C. where he knew that he would find friends in several of the Federal Scientific Bureaus. On the way there, Morris went via Philadelphia to visit some of glass works in southern New Jersey.

Outside the first works he called at he saw a sign saying—KEEP OUTSIDE. THIS MEANS YOU! But Morris went right on inside and received a very warm welcome. Regarding the manager of one of the works he visited, Morris wrote:

"He was a mixture of the best type of British skilled glass worker, and the American manager".

Morris spent 2.5 hours in the works and on leaving the manager said to him:

"You say that you are not a professional glass manufacturer, Doctor, but you seem to have some practical grip of the glass trade."

Morris was particularly interested in one specific works in which table working with glass tubing was carried out on a large scale. They were manufacturing, by mechanical methods, spinning pieces of glass tubing held in lathes, and shaping the heated end with a variety of tools. Morris looked on this in wonder; he had only begun to work in this way at Walthamstow but limited facilities had prevented progress.

He next stayed briefly in Philadelphia seeing the sights before arriving in Washington D.C. on September 21st. The next day he spent with Dr. William F. Hillebrand, a noted analytical chemist, at the U.S. Bureau of Standards. The two men were well acquainted by name in conjunction with helium and Hillebrand greeted Morris warmly. The Bureau had intended to engage in the manufacture of optical glass but the war ended before they could begin this venture. Hillebrand introduced Morris to Dr. Charles L. Parsons who was the Secretary of the American Chemical Society (ACS) and the three men had dinner at the Cosmos Club that evening.

The Midwest

Parsons persuaded Morris to travel with him to Chicago. Morris thoroughly enjoyed being there and he met many scientific men interested in glass while in Chicago [including W.E.S. Turner from Sheffield]. The keynote address of the meeting he attended dealt with glass making; Morris felt his being there was very worthwhile. When he saw a cylinder of helium from a plant operated by the U.S. Bureau of Mines, Morris asked Parsons to help him obtain some of the gas. Later, Morris received, under the authority of the American President Warren G. Harding, a cylinder of the gas, on the condition that he did not liquefy it.

Morris was also a guest at the J. Willard Gibbs Award dinner meeting of the Chicago Section of the ACS and at another dinner meeting of the American Ceramic Society of which he became a member. He also lunched with W.D. Bancroft (President of the Electrochemical Society) and J.G. Taylor, who now lived in America.

On September 27th, Morris went to Toledo, Ohio. While there, he was taken around the Libby Glass works by its manager, Hess, who Morris had met in Chicago. Morris was shown a crystal bowl two feet in diameter, which was the largest piece of cut glass ever made at that time; Morris thought this item was magnificent. He also saw the first Welsbach machine blowing electric lamp bulbs at sixty per minute and glass tubing for electric lamp manufacture being drawn, by machine, at a rate of one hundred feet per minute. Morris knew that this was not possible in England. The manager gave Morris the whole morning and then introduced him to the managers of the Libby Owens Works and the Ford Plate Glass Works where Morris spent time also. Morris left these facilities with the impression that simple visitors to such works were not welcome; management would though warmly welcome anyone like Morris who had scientific and technical knowledge and experience for honest and open discussion and sharing of ideas.

Pennsylvania, Cleveland and Niagara Falls

Morris left Toledo and traveled to Pittsburgh by night. After arriving there he met a chemistry professor named Silverman, who had once been employed in a glass works, at the university. He spent a morning there with Silverman. He also had lunch there with Libby, a Professor of the History of Science, and the two men talked about Ramsay.

He then went to the Mellon Institute. The Mellon brothers had put up $1,000,000 to start their Institute but had left the working of it to able scientists and technologists; the brothers took an interest in what happened but did not interfere. Morris knew that was what J.N. Tata would have done with the Indian Institute of Science had he not died in 1904.

Morris spent ten days visiting works in Pittsburgh, Washington, PA. and then took a night train to Cleveland to visit the Nela Park Electric Lamp Factory of the General Electric Company. He spent the whole day there with Dr. Clark (the chief chemist) and his staff. They called their Research Department the Nela University and this was not an inappropriate title. After dinner, Clark and his staff took a delighted, but exhausted Morris to the boat for his trip to Buffalo.

There followed two free days where Morris could visit Niagara, which he did as a tourist. When he arrived at the place from which one could see the falls from behind, on the Canadian side, he was given a raincoat and told where to go, which he did alone. Morris walked along a dimly lit tunnel hearing an ever-increasing thunderous roar. Then, after turning a corner, he found himself standing on a bridge, composed of a couple of planks, with a hand-rail between him a mass of green water falling from above. Morris stated that he never felt so alone in his life, quite cut off from the world. The next day he made the trip downstream to the whirlpool and in the afternoon took the voyage on the *Maid of the Mist*. Morris saw much that was beautiful and was amazed by the seamanship of the pilot who put the ship exactly where he wanted it to go.

Back to the East Coast

On October 12th Morris thought he only had one week left in America but a "long shore strike" paralyzed all Atlantic shipping. Given that he had a bit more time than he originally planned, Morris first spent a day in Rochester, New York. While there he visited the Bausch & Laumbe Optical Glass Co. where he saw the mass production of high quality goods, specifically camera parts. He spent an interesting morning there and next traveled to Corning, New York.

In Corning, he saw that they had a first class works with a large scientific staff and some very able men. To date, Morris's trip had cost him £20 a week and he had spent all that he had intended to spend. However, with no end in sight for the strike, he realized that he could still visit some university and Federal Research organizations that he now wished to contact; he was not at all worried over what he would do when he would spend his last pound.

He now traveled to Boston where he spent a morning with, and had lunch, with Theodore William Richards at Harvard University. Richards had won the 1914 Nobel Prize for his work on revising atomic weights. Morris observed that Richards though felt his work was not at all important and that he was a very, very depressed man. Morris then spent a morning at the Massachusetts Institute of Technology with Professor Henry P. Talbot and in the afternoon Morris, Talbot and Mrs. Talbot drove fifty miles over the battlefield of 1775 and the Lexington woods.

Morris had a chance to see the blazing colors of the maples and observed the white painted colonial houses with the pumpkins and red peppers set out to dry on the verandas. The next day, the Boston Sunday paper published a double page headed Illustrations of Colonial Architecture, which he took home for his architect brother Wilfred.

Next, there was a visit to the Research Division of the General Electric Company at Schenectady, where Morris stayed for a couple of days with the Director, Dr. Whiting, and his staff.

Time to go home—Not Just Yet!

Morris returned to New York City on October 22nd hoping to catch a boat home but it did not sail due still to the strike. He then went with a friend from the British War Mission to Baltimore where Morris spent a whole day seeing an up-to-date factory manufacturing the metal filaments for the finishing of the electric glow lamp. He next traveled to Washington, D.C., particularly to visit the Geophysical Laboratory though its staff had not yet been reassembled from its wartime activities.

Another visit to Pittsburgh occurred to again pay a visit to three more works. At the first works, where they were manufacturing scientific glassware, they were very good to Morris; so much so that he felt at home. The people at this works seemed to take the view that Morris had done them an honor by coming to see them from such a distance, rather than they were granting him a favor in seeing what they were doing. Each side learned much from the other in this visit; Morris learned a few tricks of craftsmanship not only from the staff but also from the employees in the works proper.

Back to England—Finally!

On November 5th, 1919, Morris began his return to England on a cargo boat departing New York City. He came home with W.E.S. Turner and one of Turner's friends. Morris shared a cabin but was as comfortable as he would have been on an ocean liner. The 20 some passengers on board held a joint service on November 11th and sang "God Save the King" and "My Country Tis of Thee", naturally to the same tune. The overall voyage home was pleasant for Morris, made more so by the fact that a large number of the passengers on board were American children.

Morris was quite puzzled as to what he would do next with his life. The glass industry in Great Britain was now in very, very bad shape and Morris wanted nothing to do with it; in fact he told friends who had invested in a new glass works in Canning town that they should get out of this investment immediately. He was right—soon their stock was worthless.

By the time Morris arrived back in England, the Department of Scientific and Industrial Research had come into existence and it was proposed to form a Research Association for the Glass Industry. This matter got into the hands of a group associated with British Glass Industries. Soon, a Director was appointed for the Research Association for the Glass Industry at a salary of £4,000 per year. This individual was supposed to have an expert knowledge of the manufacture of sheet glass. He was though not respected by leading glass technologists in America; because of this Morris did not accept an invitation to become an honorary member of the Association.

Morris though did make one contact with the Association. After the war firms making articles from glass tubing were quite troubled with a white opalescent ring, which appeared in the region just outside that in which the glass became soft in the blowpipe flame. A representative of one firm told Morris that the Glass Research Association had looked into this and concluded that the poor quality of British glass was the reason.

This made Morris wonder since the phenomenon was not new to table-worked glassware. It was tradition that all clouding of glassware in the process of manufacture was termed "sulphuring". Morris began to experiment to see if there was some element of truth to this tradition. He obtained three samples of glass tubing described as:

(a) High quality pre-war German;
(b) Duroglass Ltd. soda-potash glass and
(c) A British glass from another firm.

All three samples behaved identically in the experiments that simply consisted of heating each sample till soft and then drawing it out. The blowpipe was supplied alternatively with air or oxygen and with:

(a) Coal gas from the main;
(b) Pure hydrogen and
(c) Hydrogen, which had been passed through a wash bottle containing a little
     carbon disulfide.

These experiments showed that the clouding was only produced when the gas contained sulfur and when the flame was oxidizing.

Qualitative observations showed that the clouding took place in regions at the end of the flame. Simple thermodynamic considerations indicated that there might exist a limiting concentration of sulfur in the gas above which the concentration of sulfur trioxide (SO_3) in the gas would result in the reaction between SO_3 and Na_2SiO_3 forming sodium sulfate and free silica.

Morris gave a paper on this to the Society of Glass Technology's meeting at Newcastle on March 6th, 1921. On the next day, members of the Society visited a glass works where they manufactured heavy pressed tumblers. The worker kept the flame luminous by tossing in pieces of bituminous coal. When he was asked why he did this, he replied that it was to prevent sulfuring. This practice of the craft showed that Morris's theory was correct. It turned out that the real cause of this trouble was because less sulfur was removed from the town's glass works after the war than before the war. This work was Morris's last significant contribution to the study of glass for many years.

CHAPTER XXV

End of Glassware Work and On to Industrial Processes

Conversation with Donnan

Morris had seen very little of his friend Donnan during the war and was quite pleased to see him again after the war had ended. While exchanging war experiences, Morris found out that Donnan had been a consultant to Lord Moulton, who was Head of the Department of Explosives Supply. This conversation dealt mainly with the methods of Lord Moulton's chief assistant Kenneth B. Quinan who was an American chemical engineer with extensive explosives experience before the war in South Africa.

Donnan stressed Quinan's insistence that the development of every new chemical manufacturing process must involve the preparation of chemical and thermal balance sheets. This idea was immediately appealing to Morris. Given that partial analyses were usually more misleading than useful, which Morris had found to be the case in connection with glass, he approved entirely of Quinan's outlook. Donnan suggested that Morris should therefore try applying this method to glass making; thus it was that Morris decided to work out the chemical and thermal balances for the melting of glass.

A chemical (material) balance sheet formed, in Morris's time, a useful check on the quality of the analytical work on which the reaction was based. A thermal balance sheet gave an initial indication whether a process would 'go' or not; and it formed a basis for thermodynamic studies of the changes involved in the process, which should be done before experimental work started.

Morris began looking into this during the winter of 1919-1920 and presented a paper on it in 1921 to the Society of Glass Technology titled:—"The Heat Balance of a Plant consisting of an Air-Steam Gas Producer and a Glass Tank Furnace". This work was based not on experiment but on information obtained from the literature and from some firms. Calculation results showed that in the best contemporary practice, only 14.5% of the heat supplied to the plant was actually used in melting and finishing the glass; but in ordinary practice, the % seen was lower.

Introduction to Clark and Coal Gasification

In January 1920, Morris became the Director of a private company named Dennis Simplex Furnaces Ltd., which was formed to exploit a glass furnace design suitable for small works throughout England. Though Morris was a good enough draftsman to make his ideas clear to other professional draftsmen and builders, his drawings lacked professional finish.

To compensate, Dennis Simplex entered into an arrangement with the Richmond Gas Stove & Meter Co., Ltd. of Warrington. This firm manufactured domestic gas heating appliances and also built furnaces burning solid fuels. As a result, Morris was brought into close contact with F.W. Clark who was Head of Richmond's solid-fuel-burning furnace department. Though Clark did not have a university education, he did have a very sound knowledge of the scientific side of gas engineering through practical experience, personal study, evening classes taken and an outlook on the treatment of coal for heating a town that was quite original.

Clark believed that coal should be gasified directly by the alternate action of air and steam in a plant similar to a water-gas plant. Instead of having the calorific value of 300 British Thermal Units (B.T.U.), the product would have a calorific value of about 360 B.T.U. since it would contain the volatile constituents of the coal. It would need less oil to get this gas up to any specified value for the service of the town than would water gas made from coke and thus reduce cost. The plant used in this process would differ from the plant for the manufacture of water gas from coal in that, in the first stage of the process in which the coal was carbonized, more heat would be required than was available from the water gas produced in the second stage.

Morris's initial interest in this suggestion was to study by the method of Quinan, the carbonization of coal and the manufacture of water gas. Searching the literature noted that there was a great deal of information available related to tests on carbonizing plants; but full analyses of the fuel were very rarely recorded—the only published information dealt with moisture and ash content, volatile content and the so-called fixed carbon (the difference between the weight of the residue on heating and the weight of the ash contained in it). Though this information was satisfactory for routine control of Towns' gas, it was not satisfactory for research work.

Fortunately, Professor J.W. Cobb of Leeds had done recent work on the carbonization of coal in vertical retorts. Further good fortune had one set of tests in each identical conditions. After correcting for analytical error [Professor Cobb had delegated the analysis of the coal and the results were unfortunately inaccurate] in one set of data, the material balance sheets had values that were nearly identical for % hydrogen, which proved that the experimental work was first rate overall. This clearly demonstrated the soundness of Quinan's work. Morris now could make use of the results of two independent sets of tests for his purposes without any doubt as to their validity.

Fuel Research Board

In the mid-1920s, Morris made contact with Dr. Cecil H. Lander, Director of Fuel Research, who appointed Morris to the Fuel Research, Board. The two men got along quite well, with Lander agreeing with Morris that there was a difference in tests in plants and in planned technical research. However, the board was interested mainly in the production of high quality coke in Great Britain to which Morris gave much time though it meant a great deal of travel throughout England at his own personal expense.

Eventually, a significant amount of information was gathered and a small scale coking plant was to be built at Greenwich. Just when money was asked for from government, it was decided that the full effort of the Fuel Research Department was to be devoted to the study of the production of oil from coal. This decision followed the publication of a pamphlet titled "The Labor Party's Scheme for the Production of Oil from Coal", which was said to include the work of a group of scientists. The fuel research organization lacked sufficient staff to carry out this work; this caused Lander to resign as Director. After asking about his successor and hearing that this position would go to the next most senior member of the staff, Morris also resigned from the Fuel Research Board after having spent two years on it. The work on coke was dropped for the next 20 years as also was an investigation into the manufacture of water gas that Dr. Lander and Morris had planned.

Gasification of Coke by Air and Steam

In looking up information about the water-gas process, Morris was only able to find the results of tests, not research that had been carried out by Lander at Greenwich, at the Birmingham Gas Works by Parker and Cobb or by the Institution for Gas Engineers. Morris was himself trying to determine how the compositions and thermal properties of the blow and run gases were related to each other. He wanted to run the tests with arranged series in which:

a.   The depth of the fuel bed, the velocity of the air blast, and the time of blowing were kept constant; the blow gas would be analyzed and

b.   During this series of experiments the conditions of steaming would be varied to determine those under which the yield of water gas was most satisfactory.

However, this did not happen and Morris had no choice but to statistically treat the data, which were available and make the best of a bad situation. Ultimately, a statement was made that "the work was of little value as it did not specify exactly the kind of coke used" criticized Morris's work. Morris replied saying he had assumed at both locations (Greenwich and Birmingham) a good quality gas-coke had been used, which sufficed. In dealing with the water-gas process, Morris assumed that three sub-processes operated:

a. The blow, in which air was blown into the fuel bed.
b. The run, in which the fuel bed was steamed.
c. Clinkering, in which some coke was removed with the clinker, and some burned by air-draught through the plant.

According to the method of operating, the CO_2/CO ratio did vary. It was further true that the amount of heat supplied to the plant during the blow process was dependent on the average value of this ratio and on the quantity of carbon consumed during the blow process. This was, of course, contingent on the length of the blow process, which was limited by the fact that it resulted in raising the temperature of the coke and of the blow gas. The results of Morris's work showed that the proportion of the gas formed from coke supplied to the plant depended mainly on the CO_2/CO ratio in the blow gas and the rate of production of water gas. Morris wanted to look into this further but this work was never carried out.

Regenerative Gasification of Coal for Towns Gas

In 1921 Morris and Clark became dual partners in a private company, Travers & Clark, Ltd.; Morris had his office in Aldwych, London while Clark had his office in Deansgate, Manchester. Their work initially was primarily connected with the glass industry and they were moderately successful with this venture. A short time later Morris and Clark were pleaed that George A. Mower and George R. Thursfield, who were Directors of the Sturtevant Engineering Co. Ltd., joined them as Directors of a new company; Regenerative Coal Gasification System Limited—where Clark was Managing Director and Morris was Scientific Director.

How the process Morris developed operated is best explained by comparing it with the gas producer, and contrasting it with the water-gas plant operated on the water-gas principle with bituminous fuel. In the gas producer, when the bituminous fuel was gasified by the simultaneous action of steam and air, the coal was coked by the heat carried to it by the hot gas generated in the lower part of the fuel bed. In the region where the carbonization occurred there was little chemical change except as was involved in the destructive distillation of the coal; in this part of the plant solid surfaces became covered with a tarry layer, which prevented these surfaces from acting as proper catalysts. This became significant enough that the $CO—CO_2—C$ reaction reversed and the water-gas reaction was basically frozen.

There was no greater difficulty, according to Morris, in effecting the carbonization of the coal in a water-gas plant than in a gas producer; but while in the gas producer the products of the distillation of coal were carried forward with the producer gas, in the water-gas plant operated with bituminous fuel, only that part of the distillation gas that was carried forward by the water-gas became available while that part carried away by the blow gas was lost.

To deal with this difficulty, there were many processes available though the best known was [in England] the Tully process and the Strache process on the European continent.

British Patent 198,777

The principle for the plant devised by Travers and Clark was protected by British Patent 198,777, which was based upon the working of the gas producer that was at the time, the most efficient process for gasifying carbonaceous fuels.

The generator consisted of a carbonization chamber, which was a vessel built with solid walls and well insulated externally that was placed upon a water-gas generator [called the gasification chamber of the plant to which air and steam could be admitted intermittently]. Coal was fed from the hopper and a pipe led from the top of carbonization chamber through a cooling and purifying plant to the gasholder. The "nostrils" led into a channel, surrounded the generator, and "communicated" with the regenerator vessel, which was lined with firebrick and filled with brickwork. Also included with this vessel was a stack valve at the top for escaping blow gas. In the top of the stack valve was a pipe where gas could be drawn from the main and forced into the regenerator.

During the blow gas "phase" the stack valve was open and the valve on the pipe that connected the outlet of the carbonization chamber to the gasholder was closed. This allowed the blow gas to pass through the regenerator, to which secondary air was admitted if necessary, with the waste gas escaping through the stack valve. At the end of the blow "phase" the stack valve was opened so that gas could leave the top of the chamber. Steam was then admitted to the generator and the water gas produced passed upwards through the coal in the carbonization chamber through the outlet pipe to the scrubber and condenser and then into the gasholder.

While all this was happening, a circulator was set in motion and a small amount of the mixed water gas and coal gas was drawn from the main, forced into the top of the regenerator where this gas became heated in passing through the brick work and then entered the generator through the "nostrils" at a temperature between 800 and 900 degrees. Here the circulation gas mixed with the water gas generated in the plant, so that any desired amount of hot gas could be passed through the coal so as to effect its complete carbonization. The quantity of gas circulated did not have any relation to the rate of production of water gas in the plant. Provision was also made for down steaming though, as Morris stated, this was simple compared to the process described here.

Morris observed that though the blow gas did not pass through the coal, the heat of the blow gas was used regeneratively by transferring it to the circulation gas and passing that gas through the coal. In order to render this process efficient, it was necessary to design the plant so as to make it possible to transfer the heat of the blow gas to the circulation gas to the coal without any undue loss, and so that the temperature at the outlet from the plant should be the minimum needed for the effective removal of the tar. It was obvious that the problem was quite complicated; and though the modern gas producer was an efficient machine, the results obtained in modern practice were not achieved immediately and Travers and Clark did not expect to overcome all the difficulties connected with their process in early work.

Orders and Plant Designs

With all the plant details worked out, and in agreement with the Aylesbury Gas Company, a plant was erected at their works to produce about 300,000 ft^3 of gas in 24 hours with a calorific value of (360 B.T.U./ft^3) from coal, air and steam only. The plant, to be taken over by the company provided certain guarantees were fulfilled, was started up early in 1923. Morris was to spend the better part of the next nine months or so studying, on site, the operation of this plant. He was quite satisfied with how well the gas works were equipped for experimental work and that there was even a small laboratory near the plant.

When the plant had been in operation for several months tests on it were carried out by Dr. Pexton, from Leeds University, on behalf of the Preston Gas Company and by members of the technical staff of the Brentford Gas Company. The results of both of these tests showed that the gasification of coal by this new process was quite impressive.

In 1924, at the invitation of Samuel Tagg (the Manager of the Preston Gas Company and also the President of the Institution of Gas Engineers), Morris gave a presentation of his work with Clark at the annual meeting of the Institution of Gas Engineers in London.

Following this meeting, two orders were placed with Regenerative Coal Gasification System Limited for plants. One was for a plant of 1,000,000 ft^3/ 24 hours and it was by the Brentford Gas Company to be installed at the Harrow Gas Works facility and the other, for 500,000 ft^3 / 24 hours by the Preston Gas Company. In both plants, the calorific value of the gas was to be increased by carbureting with oil to a minimum of 450 B.T.U. /ft^3 and to 500 B.T.U./ft^3 when required.

To make the step from a small plant to the larger of these two was perhaps rash especially when it involved altering the design of the plant so as to produce a gas of high calorific value. In the Aylesbury plant oil was sprayed into the regenerator during the circulation process to allow gas of about 400 B.T.U. to be produced but since this gas had to pass through a region in the bottom of the regenerator where the temperature was far in excess of what was needed for efficient "fixing", the oil was over-cracked and the carbureting efficiency was low. Such a primitive "arrangement" could not possibly be duplicated in the larger plant; there were also engineering difficulties in the design of the larger plant that were greater than seen before. In hindsight, Morris admitted it might have been better to have constructed and tried out the 500,000 ft^3 plant before proceeding to design and then build the 1,000,000 ft^3 plant.

Even though the plant designs for the two larger facilities were similar to the previously described design, there were differences. They included: the carbureting process; changing the intake of the circulator from the gas main to a point more distant from the gas outlet; the timing of the blow and run periods and the quality of coal used in the process.

Before the plant at the Harrow Gas works had been completed, the Brentford Gas Company noted that the Harrow works had been purchased by the Gas Light and Coke Company and that they had taken over the contract between Brentford and Morris's firm. The new owners told Morris's company that they found it inconvenient to operate a single unique unit, the 1,000,000 ft^3 facility, in their large organization. They further dictated that the new plant would be paid for by them provided it passed the tests specified in the contract; whether it worked or not they were going to pull it down and replace it with standard carbureted water gas equipment.

Clark's Illness and the "Test"

In 1926, Morris had another problem to deal with. His partner, Clark, who had been suffering severely from an explosion in a gas works had a relapse and was now a very, very sick man. In August, the firm of Travers and Clark ceased to exist though Clark remained a Director of Regenerative Coal Gasification System Ltd. and promised to help Morris as much as possible with the upcoming tests on the Harrow and Preston plants. Unfortunately, just before the test at Harrow was to occur, Clark wrote to Morris and said that his health prohibited him from being of assistance. In addition, the only person that was retained from Regenerative Coal Gasification System Ltd., besides Morris, was his young assistant engineer Whitney who had worked at the Harrow facility throughout its construction.

A difficulty in carrying out the test of the plant was that the guaranteed output of the new plant was only half that of the normal output from the works. The test would also mean that the normal output of the works would be shut down and that the new plant, which had not yet been in continuous operation for 24 hours, would have to maintain quantity and quality of product during that time period. This certainly caused great anxiety to the engineer, manager and chief chemist of the facility. The Gas Light and Coke Company sent two senior technical officers to observe, a Dr. Watson and Dr. Griffiths.

The plant itself was to be operated by the Company's employees who were skilled water gas plant operators and known to Morris. Whitney, who understood the design and operation of the plant [though neither man had seen the plant in operation for more than a few hours at a time], assisted Morris. Three post-graduate students from Bristol University were keen and enthusiastic aides though not one of them had ever been inside a gas works before.

Conditions put down by the Gas Light and Coke Company were strict. Morris was to take over the plant, which had not been in operation for several months and had never run 24 hours at a time, on September 15th and start the seven-day test on September 19th. The inclined retort setting, which normally gave a significant proportion of the gas for town service was to be shut down and gas production would be from the water gas plant. Morris was surprised to find out that his plant would have to produce gas of a calorific value of about 500 B.T.U.; the normal output from the works was 440 B.T.U. Morris felt that the directors of the Gas Light and Coke Company were being very hard on him and his staff since they only had four days to instruct untrained assistants with no time allowed for "tuning up the plant", and thus ensuring that each man was fully acquainted with his duty. Morris though had faith in the design of the plant; he agreed readily to whatever conditions the Gas Light and Coke Company laid down without reservation, which certainly endeared him to their staff.

After a twelve day period of living on site, eating when one could and rarely sleeping, Morris, who admitted enjoying the experience thoroughly, and his staff finished their work in the plant. The test came to an end with the 12 P.M. to 8 A.M. shift on September 22nd and early that afternoon Morris was given the report of Watson and Griffiths.

Except for a short time on the 20th, when the plant had to shut down since the gasholders were full, it had run continuously for seven days [21 shifts]. During the first thirteen shifts the calorific value of the gas averaged 480 B.T.U. This had been necessary to maintain quality of production from the works, but as it necessitated increasing the blow period so as to maintain a high temperature in the carburetor, this was done at the expense of the run, which of course limited the amount of gas produced. In the last eight shifts, the operation was normal with daily output being 1,013,000 ft^3. It was admitted without discussion that the Regenerative Coal Gasification System Ltd. firm had fulfilled the conditions of the contract.

As Morris was finishing his conversation with Watson and Griffiths, he was told that he had an urgent telephone call. As he picked up the phone, he was informed that Clark had died. Morris, feeling neither astonished or shocked, reacted by saying what first came to mind, which was:—"I wish I could have told him that the test had been successful".

Preston Gas Works Plant

In December 1928, A.K. Collinge, who was Tagg's chief assistant, published an excellent account of work carried out at the Preston Gas Works. This plant had been running since November 1925 and had used a variety of types of coal but none were found satisfactory. Collinge wrote of the plant:

> "The chief practical advantage of a complete gasification plant over a separate retort and water-gas plant are as follows:—The capital outlay is only about three-fifths; the relative ground space is considerably less; the labour employed is in the neighbourhood of one-third; and less than half as much solid material has to be handled. The method of carbonising the coal by the circulating system is an economical success".

Morris added that the great advantage of the complete gasification plant was that it was cheaper in that it saved a good deal of oil for the gas that was made.

Tagg had anticipated a boom in trade but instead a serious recession occurred nationwide. He was therefore forced to close the whole of the works in which one plant was located.

Success in the tests of the two plants had left the Regenerative Coal Gasification System Ltd. firm just barely solvent. Clark was dead and Morris, no longer a young man, had to provide for his family. What would he do next?

CHAPTER XXVI

Return to Bristol University and Zinc Smelting

Personal Interlude and Research

The early part of the 1920's were a happy and successful time for Morris, his wife and children. The health of Dorothy and Robert was now good; however Morris's mother Anne had died in August 1923. Morris had come into some money from Anne's will and through the purchase and sale of a part of London Square he had made a profit of £1,750. Morris and Clark had also made a reasonable profit from their furnace building contracts. This money had allowed Morris to buy a home at 17 Stafford Terrace, within a hundred yards of his old home at 2 Phillimore Gardens, Kensington.

During the post-war years Morris had kept in touch with University College, London where during the years 1922-1924 Donnan had placed a research student at Morris's disposal. Morris had had an idea that the study of the expansion of gels might put additional light on the study of glasses below their softening points. Allan Taffel had done good work though it showed that gels and glasses behaved completely differently. Morris was at that time interested in the Nernst Heat Theorem and how it could be applied to technical problems. These ideas had Morris in constant contact and discussion with W.E. Garner at Bristol University.

Appointment in Bristol and Time at the University

Morris had remained in contact with Francis in Bristol over the years. In October 1926, Francis wrote to Morris saying that because of the resignation of Professor McBain (who was going to Canada), his department was being reorganized. Morris, now 54 years old, knew that he was not qualified for McBain's position, which would go to a younger man whose interest was in the rapidly expanding and developing field of physical chemistry. Francis realized this but still wanted to do something for his friend Morris. He thus proposed to the senate and council of Bristol University that Morris should be offered a lectureship in Applied Physical Chemistry at the rank of Professor. This proposal was strongly supported by the University's Vice Chancellor Dr. Thomas Loveday and was approved unanimously. Morris's friend W.E. Garner was, at these same meetings, selected for Chair of Physical Chemistry.

Morris immediately accepted the offer presented to him, dating from June 1927. He felt that the university could not have more handsomely rewarded him for the service he gave it from 1904 to 1906; he was offered and accepted a large laboratory on the ground floor where the working conditions and facilities given to him were amongst the best he had ever seen. In addition, Francis was department head with Garner second in command. Conditions, in Morris's mind, were ideal.

His duties consisted of supervising two or three students in his laboratory during the last year of their honors course in chemistry. Morris was also required to deliver some lecture courses from time to time. To undergraduates he gave lectures on the rare gases that were exciting and gave the whole story to the students, both historical and scientific, as a person close to this work. He also gave, for a few years, a graduate level course on plant design and applied chemistry but this was discontinued due to a lack of student interest.

He had, at any given time, two or three students engaged in research for the Ph.D. degree under his direction. Morris was soon able to develop a strong research group centered in on the study of the thermal decomposition of organic vapors. His group expanded and developed new methods of gas analysis that had been so successfully used in his earlier research work. He studied molecular stability by looking at thermal changes through making a complete analysis of contents of reaction vessels after specific time intervals.

Research led by Morris devised methods for introduction of accurately measured quantities of vapors into reaction vessels heated to reaction temperatures and the removal of products without loss of material or loss of time. This allowed the detailed study of the thermal decomposition of ethane, propane, tetra-methyl methane and methylamine.

Interestingly enough, in the early 1930's Morris had to, for a short time, take charge of the whole of the Department of Chemistry. Francis and Garner were both quite ill at the same time. Morris was asked by the Vice Chancellor to take charge of the whole department, which he agreed to do in this emergency situation. Fortunately, this duty was short-lived. The only other change in Morris's position was in 1934. In that year Augustus Nash endowed a fellowship in chemistry and Morris was offered this by the Vice Chancellor of the University. Morris accepted this honor, which added to his official titles and income, but did little to change his workload.

Personal Matters

Since it now was likely that Morris would remain at the university till he reached the retirement age of 65 [1937], he purchased a home at 6 Christchurch Road, near the suspension bridge, a mere 15 minute walk from the university. Morris had workmen take the wall down between the two ground floor rooms and replace it with a curtain. The large room formed became a music room for Dorothy and an area where Dorothy and Robert both danced.

Dorothy wished to earn a degree in language at Bristol University, but since she had to first matriculate she spent a year at the Redland Collegiate School. She then entered the honors school studying German and spent a year abroad in Heidelburg; she finally took a first class B.A. and shared a prize for the best arts school graduate of her year.

Robert, now 14 years old, went to school at Canford in 1927, which was a newly established public school. Though Morris was pleased with what he saw of the school, Dorothy was shocked at what she saw while visiting Robert in November. Robert was then immediately removed from Canford. Robert continued his education with six months in Geneva, a year at the big Lycee in Caen and then a year at the Muster Schule [modern school] in Frankfurt. He returned home from these experiences tri-lingual. He then worked at home with a tutor for four months in Bristol, so he could pass the Oxford local examination and thus qualify for the university; he passed this exam easily and was at the top of its honors list. He then entered University College London to study medicine but this soon changed to psychology.

After lost opportunities to pursue her music in the years from 1923 to 1926, Dorothy began to work at her music soon after the Travers family settled in Bristol, first under the direction of Frederick Moore and then Harold Samuel. In 1937, Dorothy had an audition with Solomon who said to her:—"You have music in you, but if you come to me, you must adopt my technique". Dorothy was thrilled at this possibility and agreed to his request. This entailed study for some weeks, going right back to practice on single notes and weekly lessons in London from Solomon's assistant Bryne.

At the end of six weeks, Solomon appeared at one of the lessons and said as it ended: "Now I am going to take you myself". This he did; giving Dorothy weekly lessons that often lasted more than an hour. Morris thought this a great privilege. Interestingly enough, during the Second World War, Solomon, when he broadcast from Bristol, would usually spend the afternoon practicing in Morris's home.

Zinc Smelting

Not too long after settling down in Bristol, Morris made contact with Stanley Robson, Managing Director of the National Smelting Company Limited. This firm was chiefly involved in zinc smelting and sulfuric acid manufacture and they were ready to start up horizontal zinc smelting furnaces that had been constructed during the war at Avonmouth.

Morris immediately saw that Robson was a very able man who had imagination and the ability to concentrate on a task at hand as well as working very hard on the project he was involved in. The traditional method of carrying out any process was always for him a matter of careful study where the scientific and commercial aspects could be clearly seen.

To learn about the zinc smelting process, which was new to him, Morris made several visits to the Avonmouth and Swansea smelting works. Shortly after these visits, Morris made a suggestion to Robson that the study of such processes should involve the compilation of chemical and thermal balance sheets, on the lines laid down by Moulton and Quinan. Robson readily agreed to let Morris do this.

Some of this work had already been done by the US Bureau of Mines and a preliminary account of it had been published though Morris undertook a very complete study of the process by checking and re-checking thermodynamic data. There were some aspects of this process that had similar difficulties to what Morris had experienced in the manufacture of glass; such as the destruction of the retorts where the sintered zinc ore and anthracite duff were heated from which the zinc was distilled as vapor, to be condensed as liquid in fire clay receptacles and the behavior of the furnaces. Morris enjoyed this work a great deal but was astonished when, after reading his report, Robson asked him to join his firm as a consultant; this was a position Morris was to hold for many years.

Back to America

A problem that Robson was considering was adoption of a method of manufacturing zinc, in vertical retorts similar to those used in the gas industry, which had been developed by the New Jersey Zinc Company, Inc. in America. To study this American design, Morris [and Robson who arrived later] made a short visit to America in January 1932 to see two plants where such retorts were installed and to discuss this matter with the staff of the company. While in America, Morris was also able to make some enquiries into road traffic tunnels; he was also consulting for the Sturtevans Engineering Co. Ltd, which was looking into the construction of the ventilation system for the Mersey Tunnel.

Morris left alone on January 6th for America and in spite of bad weather enjoyed the trip particularly since he found a fellow traveler who was also a friend, Dr. Claremont of Clifton, who happened to know the captain of their ship.

After arriving, Morris spent a week in preliminary enquiries and visited a vertical retort plant near Pittsburgh with a representative from the New Jersey Zinc Company. While he was there in Pittsburgh, Morris was shown the Jubilee Tunnel, which went through a ridge just outside the city by the city's engineer. This man took him, deliberately, into the building on top of the hill into which up-take and down-take ventilating shafts from the tunnel opened. This was at the end of Morris's visit and the engineer said to Morris: "What do you think of it?"

Morris replied:

> "Very pretty; but the tunnel isn't ventilating at all. This is a rush hour. The cars are acting like pistons, and pushing the air past the openings of the ventilating shafts into, and from the tunnel."

The engineer then laughed and said:—"You're right. I was wondering if you'd spot the fault." He also had time to make contact with the staff of the Holland Tunnel, under the Hudson River from Manhattan to Jersey City, which he considered a well-worked out project before Robson arrived.

Robson joined for the second week of the visit to America. They spent most of the their time with the management of the New Jersey Zinc Company at their principal works in the Leigh Valley; but Morris and Robson also did a good deal of traveling. It was obvious that much had yet to be done with the plant but the process was as successful as could be expected of such a new venture. Robson thus advised his company to take the risk of installing the plant at Avonmouth.

Manufacture of Zinc

The advantages and disadvantages of working the vertical retort process needed to be solved while the plant was running stimulated discussion on other potential methods of manufacturing zinc. It was likely that during their American visit that Morris and Robson began discussing if they could produce the metal through use of an apparatus such as a blast furnace, which Robson insisted on.

Generally speaking, heating roasted zinc ore with carbon in a blast furnace gave a vapor containing zinc, carbon monoxide and some carbon dioxide. It was clearly impossible to condense a reasonable proportion of the zinc vapor from the large volume of gas produced in the blast furnace as liquid metal through direct cooling. It had to be removed by "washing" the gas with a shower of liquid lead, which would absorb the zinc. On cooling the liquid it would separate into two layers with the upper (lighter) layer being zinc with some lead and the lower (heavier) layer being lead with some zinc. Liquid lead would need to be circulated continually throughout the apparatus to obtain pure zinc.

Some small-scale experiments gave promising results, and a considerable range of theoretical investigations were carried out but were brought to a halt with the advent of war on September 1st, 1939. Later work led to Robson building two blast furnaces that produced a total of 70 tons of zinc per day.

CHAPTER XXVII

Retirement and Death

Bristol University

Even with his "retirement" from the university in 1937, Morris still continued to be active in research on a variety of topics including: the composition of the mixture of rare gases from the hot springs at Bath; the formation of methane and condensation products by the pyrolysis of ethane and ethylene; the primary decomposition of ethane and the reaction between ethane and nitric oxide; the pyrolysis of 2-carbon and 3-carbon paraffin and olefin hydrocarbons and methods for determining the rate of chemical reactions in the gas phase. This was a significant record for a three—year period (1937-1939) let alone someone who was retired.

Amazingly enough, Morris did, after World War II, still engage in research. In 1954, when he was *82 years old*, his last known research work appeared in the Proceedings of the Royal Society; this effort looked at the thermal decomposition of the lower paraffin hydrocarbons, of paraffin-olefin-hydrogen equilibrium mixtures and of similar compounds and systems.

He also did not agree with the accepted view that hydrocarbon decomposition began with free radicals and was carried on by a chain reaction. Morris instead believed that his experimental results allowed the simultaneous occurrence of two different processes: one which led to the hydrogen and olefin only, which may have involved free radicals and a second step that did not involve free radicals which he considered much more important. These views conflicted with those of many of his contemporaries.

Work during World War II

With war beginning on September 1st, 1939, Morris soon joined a research group that included W.E. Garner and E.L. Hirst, which was working for the Armament Research Department of the Ministry of Supply. He also found himself a member of a committee that was formed to consider how to provide for the large number of light bulbs needed to meet the demand of factories working by artificial light for 24-hour days [given the black out through the country]. After some discussion, even including the folly of hand manufacture of such bulbs and the damage that would occur from a German bombing raid, it was decided that the real solution was to buy American machinery and make the glass in modified bottle works. Morris, at the age of 68, also became a consultant to the Explosives Section of the Ministry, which he continued to do until the war ended in August 1945.

He traveled extensively throughout the country during the war and was commonly seen with his thermos of coffee and foot warmers. Specifically, his work involved visiting explosives factories and advising on technical problems dealing with making explosives and propellants. Morris also designed and helped to develop plants manufacturing sulfuric acid; he also advised on the fuel supplies needed by the munitions factories. His responsibilities also had him acting as the Chairman of the Cordite Manufacture Committee.

Indian Institute of Science (1940's and 1950s)

From 1942 onwards, the Indian Institute of Science [now under the able leadership of its 5th Director J.C. Ghosh] began new courses of instruction and branched into research areas in such subjects as aeronautical engineering, metallurgy, internal combustion engineering and power engineering. It also added many new buildings and laboratories to house these new departments.

When India achieved its independence in 1947, Morris was officially informed that he would have to pay Indian income tax on his pension. He replied back saying that it had been agreed upon that his pension was paid out in sterling, in London, with no deductions at all. This was good forward thinking on Morris's part; he did not believe that in 1914 that British rule in India would last even another 30 years. The outcome of the pension tax question though led to Morris begin a correspondence with the Registrar of the Institute, A.G. Pai that saw many friendly letters exchanged between the men in the next few years.

Further correspondence that Morris had with India showed that Institute faculty felt they still had a way to go to achieve the status of institutions such as the Carnegie Institute or the Massachusetts Institute of Technology; but they were at least on the way to doing so. It was noted to Morris that if he were able to visit the Institute, impossible due to his age and health, he would be quite pleased with what he had carried from idea and brought into reality.

However, a disheartening feature of the Institute that had developed since the time India became independent was in the tendency to centralize authority and control in the central office of the Institute. This led to the Institute adopting rules and practices that were prevalent in Indian government. Of course, this was viewed as poor but staff members could only do what they were told to do.

Golden Jubilee of the Institute (1959)

In February 1959, there were elaborate celebrations to commemorate the 50th anniversary of the Indian Institute of Science. This event, held in Bangalore, had several British ministers and the Duke of Edinburgh present. Though Morris was invited to attend, he could not do so—he was after all 87 years old at this time. He though did send a message of congratulations and good will to S. Bhagavantum [the 7th Director of the Institute and a well-known physicist]. Morris though received the honor of contributing to the All India Broadcast on the founding of the Institute, which he did in the following words:

> "That the Indian Institute of Science has been established in Bangalore, where conditions for study and research are ideal, is due to two great Indians. In the first place to Mr. Jamsetji Tata, who conceived the idea of bringing science and technology to the aid of Indian arts and industries and decided to set aside a portion of his fortune for this purpose. His friend Sir Sheshadri Iyer, a man of similar outlook, at the time Prime Minister of Mysore State, at once promised financial and material support for the scheme if the Institute were established in Bangalore.
>
> The whole idea was new to India, and various schemes for an Institute were put before Mr. Tata, but being a good judge of men, he at last decided to accept advice from Sir William Ramsay, Sir Orme Masson, and Sir Charles Martin, than whom no abler advisers could have been found. He put his final views in a letter dated April 8th, 1904. He died six weeks later.
>
> I was nominated first Director of the Institute by the Royal Society, and was appointed by the Secretary of State for India, with instruction to develop and carry out the conditions laid down in the late Mr. Tata's letter of April 8th, 1904. I arrived in India on November 22nd, 1906, and spent eighteen months traveling about the country and studying the problems with which I had to deal. I had to meet difficulties, arising out of the fact that ideas on university education in Great Britain had undergone fundamental changes during the previous ten years, and these ideas had not reached India. Bur all worthwhile problems are difficult.
>
> In 1909 I put before a committee consisting of representatives of the government of India, the Mysore government, and Mr. Tata's executors a scheme for the Institute, which I described as being the nucleus of an institution like the Imperial College of Science and Technology in

London. It was of university status, consisting of independent and yet inter-dependent departments. A great library, and departments coming under the headings chemistry, engineering and physics, and biological science represented at first by applied bacteriology, were to be established first of all.

The scheme was adopted by a meeting of a Council of the Institute in 1909. Buildings were erected, and the Institute was open to students in July 1911. When I India in June 1914, I felt satisfied that the great experiment had succeeded beyond my own expectations.

I claim only to have given India a plan and a policy according with the latest Western ideas.

I would have been delighted to have been able to visit India and to see how others had built on the foundation I had laid. I don't think that I would have been disappointed."

This broadcast was heard well in England but it was difficult to hear in India on the opening day of the celebration. Morris was highly praised for his work by a variety of individuals. He read of the celebration in the Times of India, daily issues of which he still obtained by airmail though he felt that the real celebration should have occurred on April 8th [the day of J.N. Tata's letter of 1904].

Biography of Sir William Ramsay

In February 1918, approximately 18 months after Sir William Ramsay had died from nasal cancer; William Augustus Tilden wrote a biography of Ramsay titled "Memoir of Sir William Ramsay, K.C.B., F.R.S." Though the author was well aware of Ramsay's scientific contemporaries, he was not though a close friend of Ramsay. The contact that Ramsay and Tilden had was only at formal social events or meetings of the Royal Society and/or Chemical Society.

Tilden, in his work, had contacted and received information on Ramsay from many individuals, including Ramsay's students. These people were only acquaintances of Ramsay and while the book was being written World War I was raging; many of these same individuals were engaged in war work and could only answer Tilden by letter. Though Lady Ramsay helped Tilden in his work with the chapter she composed titled "Notes on Travel" and through the loan of a number of letters, Tilden, perhaps due to no secretarial assistance, embodied many original documents in his manuscript, which was unfortunately destroyed after Tilden's book was published.

After the death of her son, William George Ramsay in 1927, Morris was asked by Lady Ramsay to look through Sir William's papers and books in the laboratory and the library of her house at High Wycombe, which she was moving out of. Fortunately, Morris found a collection of laboratory notebooks which had a record of the work done on the discovery of the rare gases and of early work on radium. Lady Ramsay, though hesitant to loan more of her husband's materials to anyone, let Morris have them since he was a close personal friend. Morris thus used these materials to write his second book, The Discovery of the Rare Gases, which was published by Edward Arnold & Co. in 1928.

When Lady Ramsay passed away in 1937, she left behind a large number of carefully arranged documents and letters from her husband that she obviously intended to preserve. Her daughter Catherine assured Morris that this had been done [by Lady Ramsay] in the hope that it would help to someday write a more personalized biography of her husband. Catherine soon asked Morris to take over the whole collection and sort out those papers that which seemed most worthy of preserving.

This effort of Morris's came after his retirement from the University of Bristol and occupied a good deal of his time from 1937 to 1939 [which was also the time period that Morris was President of the Faraday Society] but was not completed when war began on September 1st, 1939. At this time, Morris put all these papers in his bank in Clifton where they remained safe until World War II ended in 1945. Morris was only able to resume work on this task in 1948 when he was 76 years old. At this time, he added to a considerable number of papers he had collected; including many letters from friends of Sir William who had been associated with his scientific work.

This collection of papers was then bound in 24 volumes, with introductory notes covering successive periods. In 1952, Morris (now 80) with Catherine offered to present the Ramsay papers to the University of London for preservation in the library at University College London on the Centenary Celebration of Ramsay's birth. Their offer was enthusiastically accepted.

The following year, 1953, Morris began writing his biography of Ramsay. This work took two years and involved significant assistance from Morris's wife Dorothy who gave up her music to be of help. It also involved assistance from Dr. Clifford G. Silcocks who helped read and critique Morris's work and from Dr. Francis L. Usher who helped in making corrections, indexing material and seeing the book through its galley and page-proof phase after Morris suffered a serious accident in 1955.

Edward Arnold & Co. finally published the final work, A Life of Sir William Ramsay K.C.B.—F.R.S., in 1956. This book attempted: an analytical study of Ramsay's work in both science and education; to trace the development of his ideas; and of the experimental methods that he used. Morris also, unlike Tilden, tried to draw a picture of the man whom Morris and others knew and loved.

Autobiography and Death

Having authored three significant books The Experimental Study of Gases (1905), The Discovery of the Rare Gases (1928) and A Life of Sir William Ramsay

K.C.B.—F.R.S. (1956), Morris now began working on his own autobiography. This task, largely accomplished while he was in his mid to late 80's, is what has been edited and updated in this work. Additional information, not originally noted by Professor Travers, has been gathered by the primary author and has been graciously supplied by members of the Travers family.

Morris William Travers, F.R.S. passed away, after a long illness, on August 25th, 1961 in his home at Stroud, Gloucesterhire where he had lived since departing Bristol in 1949.

CHAPTER XXVIII

Recollections of his son

Before concluding this biography, largely based on the unpublished autobiography of Morris William Travers and other items researched by this author, it is interesting to look at the perspective of his son, Robert Morris William Travers (1913-2004) who was a well-known educational psychologist who worked for, in part, the United States Air Force, the University of Utah and Western Michigan University. Following are observations on his mother and on his father, Morris William Travers.

His Mother Dorothy

Dorothy Gray (1884-1970) was a short, plump, physically dynamic, expressive and outgoing individual. She and her sister, Maye, were raised to do nothing more than engage in educated conversation, marriage, polite social life and the bearing, but not rearing, of children. Dorothy and Maye both studied music in Paris; Maye actually became a professional musician while Dorothy pursued her music later in life.

Dorothy had the perception, near the end of the first decade of the 20th century, that she, by not yet being married, was an embarrassment to her parents and might be on the way to being looked at as a spinster.

Morris, on leave from India in 1909, had decided previously that he would return to India with a wife who could perform all the social functions that he had difficulty with. It was nearly miraculous that he saw in Dorothy all the qualities that he lacked and would make for an acceptable wife. She was an entertaining conversationalist who could speak about music and literature [subjects that were simply not to Morris's liking]. Dorothy spoke fluent French and German, which would prove advantageous in scientific circles. She also had a marvelous sense of humor, a great capacity to empathize with others, a natural talent for acting and could tell a story in the most refreshing way.

Morris was the exact opposite in that he had a magnificent, cold, logical intellect; he was well read in the sciences but had little use for great literature. He could talk well but was a poor listener.

They soon became engaged and were married. This marriage was an arrangement of convenience for both; Morris "acquired" a wife who had the social skills he lacked and Dorothy was given an escape from her life in England. Morris though did not know how to handle a tender, intimate relationship and Dorothy, no doubt, felt that their marriage lacked certain essential human values [she in fact, in her old age, often said that she looked longingly at the young British officers from the military station near to where they lived in India].

However, strange circumstances often draw couples closer together and Morris's political troubles in India significantly strengthened the bond with his wife. Morris regarded Dorothy as his confidant and this relationship with Dorothy enriched his life and made his problems endurable.

Morris William Travers—The Man

Though Morris was tall (six foot) and strong he was not well coordinated as a young man and thus had no success at sports. He also had few, if any, friends while a youth and thus became socially isolated very easily. He yearned for more friends but school gave him little opportunity to develop friendships. He was unhappy and was plagued with insomnia.

This isolation led to difficulties in self-expression that made many think he was a cold and austere figure. He never did understand how to give and take in a conversation, knew nothing of small talk or how to relate to an entire group. Groups frightened him and when he had to address an audience [many times of course in his career], he was literally frozen with fear.

After some negative personal experiences between the men Morris's contact with Sir William Ramsay gave him the warmth of fellowship that he had never known before. This association began to fill the vacuum that was Morris's social education. Ramsay became Morris's surrogate father and role model. Their relationship began to melt some of the frozen layers in Morris's personality; though this "thaw" was only partial it encouraged Morris to, for the first time, develop a relationship with a woman.

This young woman was a lady of wealth, natural poise and distinction. Morris pursued this romance for several months to the point that he had little control over his emotions. Ultimately, this lady suddenly terminated their relationship and was certainly appalled by her prospective fiance's awkward social behavior that bordered on boorishness.

Morris was devastated by this action. He withdrew from the world and began working 18-hour days in the laboratory. He told his associates that he was too busy to talk or even go with them to the local tavern for a "pint of bitters" after work. He slept very little and ate just enough to keep him alive but not enough to keep him from becoming thin and haggard in appearance. This continued for several months until he began taking barbiturates that helped him to sleep and also stimulate his appetite. It though took Morris more than a year to recover from the end of this relationship.

Morris had one character trait that was to make his life in India exceptionally difficult: he could not compromise. He had accomplished more by the age of 34 (in 1906) than most people accomplish in a lifetime and he had every reason to believe that his life would continue along the beautiful path it had travelled down the last 12 years. He probably felt that he could conquer all of life's problems. His limited ability to form warm and supportive relationships also did hurt his work to establish the Indian Institute of Science. He was used to having his way and was quite insensitive to the powerful political forces arrayed against him.

Morris had a great deal of information that he conveyed to his children on a variety of scientific topics either through demonstration or through "lecturing". His mind was perhaps designed like that of a computer and was ready to dispense information at the slightest cue. He though had little capacity for dialog and did not listen well to questions and would become angry rather than admit he was wrong on some topic.

His son Robert never found out what Morris's religious beliefs were, whether he thought there was life after death or whether he believed there was a God and, if so, what type of God existed. Morris went to church because that was one of the duties that a middle class gentleman "performed" in England.

Despite his aloofness, Morris was deeply involved in his children's success. He was hurt by their disappointments and celebrated in their triumphs. He wanted his children to be successful in terms of his own concept of success, which rested on some type of professional achievement.

Morris had an enthusiasm for whatever he was involved in whether it was his research, building a University or Institute, writing letters or his hobbies such as fishing. These interests remained undiminished after his retirement and until his death as was seen in his interest in research on gaseous reactions; in particular in the work of one of his former research students, Dr. Clifford G. Silcocks who was Department Head of Chemistry at Gloucester Technical College which was not far from Morris's home at Stroud.

He was a great letter writer and he kept all his colleagues, students and friends well informed of his activities and research interests. He always sent out a Christmas card to many that was paired with a lengthy letter.

He was happiest in carrying out experiments; Morris was an excellent glassblower and had significant talent in the workshop including the ability to use most machine shop tools. This was also something he expected out of all his research students; they had to be able to make their own glassware (when needed) and handle most other tasks routinely relegated to machine shop staff. Under Morris's direction all his students became capable practical laboratory workers. He could do anything he asked of his students and was always ready to demonstrate a technique or conduct an experiment himself.

He found it difficult to relax; he was always on the go in the laboratory, University or Institute tending to his experimental work or developing ideas or assisting others.

Every new result was a discovery and needed to be discussed with his colleagues. He was quite fond of telling stories of his early days at University College; specifically of his work with Ramsay on the rare gases—Morris idolized Ramsay.

Morris was also a very capable administrator; he liked things done correctly and quickly—he grew angry at inefficiency and he often typed his own letters and publications. He had the rare ability to quickly put his finger on the crux of a problem and then devise ways to solve it.

Morris, outside of the laboratory or office, had his interests in family, music and hobbies such as fishing and hunting. He also traveled widely, even before his time in India, spending much time as a young man mountain climbing in Europe. He had a wide circle of friends in England, Europe, America and in India and he entertained guests often in his home.

Edwards Brothers,Inc!
Thorofare, NJ 08086
22 March, 2011
BA2011081